MASONRY

This volume is part of a series offering homeowners
detailed instructions on repairs, construction
and improvements they can undertake themselves.

HOME REPAIR
AND IMPROVEMENT

MASONRY

BY THE EDITORS OF
TIME-LIFE BOOKS

TIME-LIFE BOOKS
ALEXANDRIA, VIRGINIA

THE CONSULTANTS: Lelland L. Gallup is Assistant Professor of Housing and Design at New York State College of Human Ecology, Cornell University, Ithaca, New York. He is responsible for a series of innovative home-maintenance courses given to New York State homeowners and home associations.

Harris Mitchell, special consultant for Canada, has worked in the field of home repair and improvement since 1950. He is Homes editor of *Today* magazine, writes a syndicated newspaper column, ''You Wanted to Know,'' and is the author of a number of books on home improvement.

Sidney F. Borg is Professor and Chairman of Civil Engineering at Stevens Institute of Technology, Hoboken, New Jersey. He has served as a consultant to federal and state governments, and private industry.

Guy Alland, an architect and the founder of the Know-How Workshop in New York City, teaches home repair and improvement. He co-authored *Know-How*, a home-repair book.

Robert Boyce, a professional mason experienced in all phases of building, demonstrated firsthand the techniques involved in working with masonry.

Louis Potts, a practical master of masonry and carpentry, has been engaged in construction projects for more than 35 years.

Joseph D'Angelo, a stonemason, has operated his own contracting firm since 1946.

For information about any Time-Life book, please write:
Reader Information
Time-Life Books
541 North Fairbanks Court
Chicago, Illinois 60611

Contents

1 Materials of Lasting Beauty   **7**

Basic Techniques of Working with Mortar and Trowel   12

The Professional's Tricks for Simple Repairs   16

How to Restore Concrete   20

Renovating Damaged Stucco Walls   24

First Aid for Holes in Asphalt   25

Homemade Concrete in Convenient Batches   26

Solid Bases for Posts and Poles   30

Making Holes and Securing Fasteners in Masonry   34

2 The Magnificent Mud   **39**

Building with Concrete: The Preliminaries   40

Pouring and Finishing   50

For a Large Slab: Reinforcement   58

A Driveway That Lasts   62

Sturdy Footings for Strong Walls   65

Rugged Steps of Poured Concrete   68

Creating Free-Form Shapes: A Bowl for a Pool   70

3 A Cornucopia of Blocks   **75**

Modern Bricks: 10,000 Shapes, Sizes, Textures   76

Classic Paths to Modern Paving   80

The Basics of Building with Blocks: A Brick Wall   86

Veneering a Concrete Slab with Ceramic Tiles   96

The Roughhewn Appeal of Natural Stone   102

The Tricks of Making a New England Dry Wall   106

Hollow Blocks for Economical Construction   112

Barbecues of Blocks and Bricks   118

Credits and Acknowledgments   **124**

Index/Glossary   **125**

Materials of Lasting Beauty

1

Building for keeps. A mason shapes mortar with a bricklayer's tool called a jointer to put the polishing touch on a wall. The wall's rugged beauty —like that of all paving and structures that are made of masonry—should last for generations without any further attention.

For transforming a backyard into an outdoor living room, nothing beats masonry. A brick wall, a stone wall, a flight of concrete steps —all are beautiful, affordable and durable. As the pictures beginning on page 64A demonstrate, masonry can enhance every sort of setting. Bricks, blocks, stones and tiles come in more colors, textures, shapes and sizes than any other kind of building material; concrete can be formed and finished to suit any design and any dimensions you can imagine. And while the bricks, blocks and tiles range from budget-priced to expensive, the most adaptable form of masonry, concrete, is generally reckoned the world's lowest-cost building material—next to fieldstones, which in many places are free to anyone who wants to take the trouble to collect them and haul them away.

Above all, though, masonry is permanent. The materials require little, if any, attention to stay in perfect condition once they are installed, and they are easy to repair or restore when they do suffer damage. They are fireproof, rustproof, pestproof and more resistant to traffic, sun, salt, spray and air pollutants than most substances—including their major rivals, wood and metal.

Masonry structures are exceptionally strong, able to bear the great weight that presses inward on them. While they are not so sturdy against stresses that tend to bend or stretch them, this weakness is easily remedied by various forms of metal reinforcement. Steel mesh gives concrete slabs enough strength to serve as driveways, and steel rods provide bonding for the concrete in heavy footings. Lightweight footings, designed to hold low walls or barbecue fireplaces, do not usually require any reinforcing.

The only major enemies masonry has are earth and ice. Sandy underpinnings, unstable backfill, earthquake zones and marshlands create unusual problems that only professional masons are able to solve. The danger of frost heaves in cold climates is also very real, but in most places an amateur can build safely just by using sand or gravel to provide a drainage bed for water, thus greatly reducing the damage that is caused by the expansion and contraction of freezing and thawing ice.

For all masonry's advantages, amateurs often avoid it because it looks like hard work. Not true. Some materials, of course, are heavy; a fieldstone may weigh 150 pounds or more, a standard bag of the cement used to make concrete weighs 94 pounds. But most materials are reasonably light, and even the heavy ones are simpler to handle with proper equipment, such as sturdy wheelbarrows and crowbars, and sensible methods, such as remembering to use your legs—not your back—for lifting. In fact, the main requirements for such jobs as laying and aligning the bricks for a garden wall are patience and precision—exactly the same requirements that you must meet to do a good job of carpentry or painting.

The masonry techniques involved in the projects in this book take only a bit of practice to master. Most of them can be learned while you are making simple repairs to existing brickwork, concrete or asphalt (pages 16-25).

Some procedures—throwing a mortar line, for example—may seem to be tricky. But this technique boils down to nothing more difficult than twisting a trowel in such a way that mortar will spill off it to form an even ridge; the knack is easy for most people to acquire.

With new construction, in fact, the biggest part of a masonry job may well be the preparation. Most projects involve excavations, although deep ones are seldom needed. For most concrete work you have to build wooden forms. And many undertakings require a building permit, which may call for special materials or techniques. It is particularly important to check the local building regulations whenever you plan to build anything that will cross or abut public land (as a sidewalk or driveway often must), or that will lie on or close to a neighboring property line, or that will represent a potential safety hazard (as a high wall may).

Once the advance work has been accomplished, the job can be very simple. Much masonry is built of blocks that go together like the playthings that children use. Bricks, tiles, and concrete blocks or cinder blocks come in standardized sizes—and in a wide variety of shapes, so that they can readily be fitted into the design of your choice. Tiles even come glued onto sheets, so that a whole sheet can be laid at one time.

Beyond the simplicity introduced when modular units are employed, there is another great advantage to masonry: speed. The actual work goes fast because cement, the basic ingredient in nearly all masonry, dries and sets at up to rocklike hardness in a matter of minutes or hours. Cement, which is formulated from limestone and clay, is a water-soluble gluing agent that holds sand together to form mortar or grout. Mortar and grout in turn bond bricks, blocks, tiles and stones into sturdy, solid structure.

When hydrated lime is added to mortar, it becomes the veneering material that is known as stucco. When gravel or crushed stones are added to a mixture of cement, water and sand, the result is concrete. There is one masonry material that is made without cement: asphalt, which substitutes for cement a crude-oil extract that also acts as a bonding agent for sand or gravel and dries almost as quickly as cement.

The speed with which these bonding agents set demands timing. That problem can be dealt with in the planning. To take the pressure off, and to put pleasure into doing a job carefully and expertly, most masonry described in this book can be planned in stages—and completed at a leisurely pace, one stage at a time.

Words Given a Craftsman's Twist

The mason's speech may sound like English but to the uninitiated it could as well be Martian, for common words have been given uncommon meanings, many only fancifully related to ordinary usage. In a few instances the word-stealing has gone the other way, and a mason's term has acquired a new definition in general speech—the traditional headgear of the university scholar, for instance, is almost certainly named after the similar-looking square board used by masons to hold mortar. Among the more colorful masonry jargon:

BAT: A segment of a brick that has been divided into pieces.

BATTER: Slope of the face of a wall.

BUTTER: A barely pourable mixture of cement and water that anchors flagstones in mortar. Used as a verb, butter means to spread mortar onto a brick before setting it in place.

DARBY: A kind of trowel used to compact and level poured concrete.

FLOAT: A board used to smooth the surface of a concrete slab. An outsized one for large slabs is called a bull float.

GREEN: Describes newly thrown mortar that has set but not fully hardened.

HAWK: A small, short-handled wooden tray for holding mortar. The term may come from the board from which peddlers hawked their wares.

HOD: A long-handled, V-shaped trough used for carrying bricks or mortar.

MUD: Wet concrete or wet mortar. The term probably derives from the earliest mortars, which were mud or clay. Walls or pairings made with mortar are still called mud-set, and those constructed without mortar are called dry-set.

POINT: To repair mortar joints by removing crumbled mortar and filling the gaps with new mortar.

RAKE: To scrape out green mortar from between bricks for decoration.

REBAR: A reinforcing rod.

SAILOR: A brick that stands upright rather than lying on its side. A soldier also is a brick that stands upright. The difference is that sailors stand side by side, broad faces showing, while soldiers line up in single file, narrow faces showing.

SCREED: A long board used to smooth the surface of freshly poured concrete, or a gravel or sand bed.

SOUP: Runny concrete.

STORY POLE: A homemade gauge used as a guide to the height of each course of brick or block.

WORKS BRICK: A brick layout that needs no block cutting.

YARD: Short for cubic yard.

Building Codes: An Ancient Tradition

Some homeowners may regard building codes as a modern invention of professionals intent on making the amateur's job harder than it need be. But codes are ancient safety precautions, and the old ones were tougher than the ones we have today.

The earliest recorded codes, promulgated by the Babylonian King Hammurabi in the 18th Century B.C., decreed: "If a builder has constructed a house for a man, and his work is not strong, and if the house he had built falls in and kills the householder, that builder shall be slain." In Polynesia, an even sterner law required builders to entomb a live slave under each corner post, which was meant to guarantee that the structure would be supported properly, in perpetuity.

Tool Kit for Masonry

Masonry work requires general-purpose tools as well as some specialized ones that are used only with specific materials —concrete, brick, stone, block or tile. All are widely available at hardware stores, and some rarely used ones—a stud driver (often called a stud gun), for instance —can be rented.

The most important general tools in any masonry kit are a 4-foot-long mason's level (or a long carpenter's level), a steel square, a 50-foot measuring tape, a cold chisel for chipping mortar and concrete and a 4-pound mallet, or maul.

Special tools needed for concrete work include a hoe for mixing mortar; a small wood float for leveling; a square-end steel trowel for smoother finishing; a darby—a long narrow float—for large areas; a fluted rub brick; a step edger; an edger for rounding slab sides; and a jointer to cut grooves in freshly laid concrete.

To lay a true course of brick or block, a mason's line and blocks are essential. Other tools for laying and finishing a brick or block wall are a bricklayer's hammer and a chisel, called a brickset, for cutting; a small mortarboard, called a hawk, to hold mortar close to the job; a mason's trowel for spreading the mortar; a joint filler for filling extra-long joints with mortar; a pointing trowel to form weathered, struck, flush and extruded joints; a V-jointer and convex jointer; and a raking tool to form a recessed joint.

A carbide-tipped masonry bit in an electric drill is fine for putting small holes in cinder or concrete block. For holes larger than an inch use a star drill, a kind of chisel meant to be struck with a hammer. When laying bricks on a horizontal surface—a wall or patio, for example—use a V-notched square trowel to score the bed joint. For most stonework, bricklayer's tools will serve. However, you may find a stonemason's hammer and chisel useful for cutting.

For tile work you will need a steel trowel that has ¼-by-⅜-inch rectangular notches to score the bed joint for quarry tile, or a box notch trowel with ¼-inch-square notches for mosaic tile. A quarry-tile cutter cuts tile to size; a tile nipper removes smaller pieces.

MASON'S LEVEL

STEEL SQUARE

COLD CHISEL

MALLET

50' TAPE

STEP EDGER

MORTAR HOE

EDGER

JOINTER

DARBY

WOOD FLOAT

RUB BRICK

SQUARE TROWEL

STONE CHISEL

BRICKLAYER'S HAMMER

BRICKSET

STAR DRILL

MASONRY BIT

STONEMASON'S HAMMER

MASON'S LINE AND BLOCKS

JOINT FILLER

POINTING TROWEL

MASON'S TROWEL

RAKING TOOL

CONVEX JOINTER

V-JOINTER

HAWK

NOTCHED TROWEL

QUARRY-TILE TROWEL

QUARRY-TILE CUTTER

TILE NIPPER

BOX NOTCH TROWEL

Basic Techniques of Working with Mortar and Trowel

Mortar is the basic bonding material that holds bricks, stones and blocks together. It also serves as a patching compound for concrete and, thinned down, becomes a grout for brickwork repairs. In all these jobs, a trowel is the essential tool. It is used to slice a batch of mortar into usable portions, to form the mortar into a foundation bed—or line—for bricks and blocks, to spread mortar on them and to shape the joints between them.

Once you master mixing and troweling mortar, you can fix walls or paving, or launch more ambitious projects such as installing a brick wall or laying a flagstone walk. The standard techniques are as useful for a five-minute repair job as for an elaborate masonry structure.

For most purposes, you can produce workable mortar by following either of the two recipes at right. They both contain portland cement as a bonding agent, sand to give strength and hydrated lime for pliability while the mixture is wet. The first recipe is frequently called a 1:1:6 mix. (The numbers refer to the ratio of the portland cement, lime and sand, and the ratio remains constant whatever the size of the mortar batch.)

The second recipe, a 1:3 mix, calls for a portland cement-lime mixture known as masonry cement to which you add the sand. You can also get the ingredients premixed—a dry mortar sold under several brand names—to which only water need be added. The ready-mixes are more expensive than ingredients purchased separately, but the cost may not matter when only small amounts are needed. Ready-mixed mortar comes in bags as small as 5 pounds, portland cement in 94-pound bags, masonry cement in 70- and 80-pound bags, and lime in 50-pound bags.

Sand for mortar can be any clean, dry, finely graded building sand. Generally it is sold in 50- and 60-pound bags. Never use beach sand; it contains salts that will weaken and discolor the mortar and prevent it from drying properly.

The exact amount of water required for the mortar depends on the humidity and temperature as well as the moisture in the sand and cannot be computed in advance. Add the water slowly in small amounts, stirring until it is all absorbed. But add as much water as possible without ruining the desired consistency. If the mortar is too wet, it will run out between joints; if it is too dry, it will not form a really tight bond.

While using the mortar, stir it often. If evaporation dries it out, add water from time to time—a process called retempering—to restore its workability. Mortar that starts to set before it can be used should be discarded. After the mortar has set —but before it hardens, usually within one or two hours—finish the joints by compacting and shaping the mortar between the bricks with a trowel or special finishing tool. Mortar only when the temperature is above 35°.

The recipes given here make 1 cubic foot—about what you need to lay 25 to 30 bricks and as much as you are likely to use before the mix hardens. The quantities are specified in gallons—easy to measure with a marked pail—although materials are sold by the pound. For ease in estimating your needs, weights are also listed. When mixing mortar, wear leather-palmed work gloves to protect your hands from irritants in the cement; keep them on when buttering and laying bricks as well. But such finishing jobs as shaping mortar joints are best done barehanded; you will get better control over the finishing tool, which will keep your hands safely away from mortar.

To make mortar decorative as well as structural, it can be colored—to accent joints between bricks or to help disguise and beautify repairs. Masonry cement comes in a variety of premixed colors, but pigments are available for creating shades in any mortar. Mix colors with the dry ingredients and, after adding water, stir until there are no streaks. Make a sample batch and let it set to see the final color; it lightens as the mortar dries. Colored mortar cannot be retempered without altering the shade.

Mixing and Testing Recipes

Ingredients to make 1 cubic foot of portland cement mortar:	Ingredients to make 1 cubic foot of masonry cement mortar:
1¼ U.S. gallons of portland cement (16 pounds)	2½ U.S. gallons of masonry cement (31 pounds)
1¼ U.S. gallons of hydrated lime (8⅓ pounds)	
7½ U.S. gallons of dry sand (100 pounds)	7½ U.S. gallons of dry sand (100 pounds)

Making a batch of mortar. Measure cement. sand—and lime if required—into a wheelbarrow. Mix with a hoe, push the mixture to one end and pour 2½ to 3 gallons of water into the other end. Hoe the dry ingredients into the water. Working back and forth, add more water and repeat until the mortar has the consistency of soft mush and all lumps are eliminated.

Generally at least 4 to 5 gallons of water or so are required for a cubic foot. But be prepared to make adjustments as you test the consistency of the mix. To do this, make a curved furrow across the surface with a hoe. If the sides of the furrow stay in place and the clinging mortar shakes freely off the hoe, the mix is ready. If the sides of the furrow collapse, the mix is too wet—add small, proportional amounts of the dry materials. If the mortar does not shake freely off the hoe, it is too dry—add very small amounts of water, testing as before.

Making a Bed of Mortar

1 Cutting off a slice of mortar. Take enough mixed mortar from the mixing container to form a mound in the center of a mortar board or a 1-foot-square piece of plywood. To separate a slice of mortar from the mound, grasp the trowel handle between thumb and forefinger, with thumb resting forward along the top and fingers curled around the handle. With hand, wrist and forearm relaxed, drop the trowel edge.

2 Loading the trowel. With a twist of your wrist, sweep the trowel blade under the slice and scoop the mortar onto one side of the blade so that a wedge of mortar, called a windrow, lies on this part of the blade. Shake the trowel sharply downward to firm the loaded mortar.

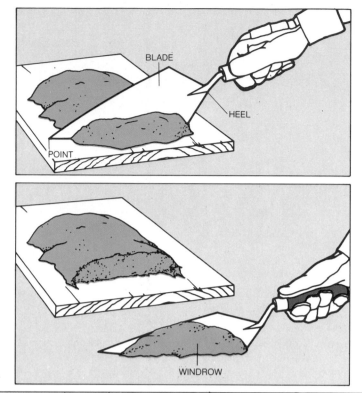

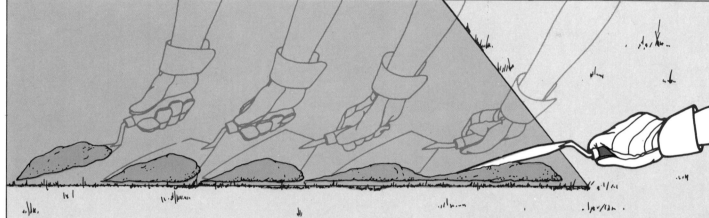

3 "Throwing" the mortar. Set the point of the trowel, face up, where you want to begin the mortar line (left in the drawing above). Bring the trowel toward you and at the same time rotate the blade 180°: the mortar will roll off and form a straight line, about one brick wide and several bricks long. The mortar line should be about 1 inch thick. If the line is not straight, return the mortar to the board and repeat the throwing action. With practice you should be able to form a mortar line four bricks long.

4 Furrowing the mortar. Immediately after throwing the mortar line, run the point of the trowel down the center of the line to make a shallow furrow. This small depression helps form a steady bond because it spreads the mortar outward slightly so that when the brick is pressed down on it the mortar will be evenly distributed.

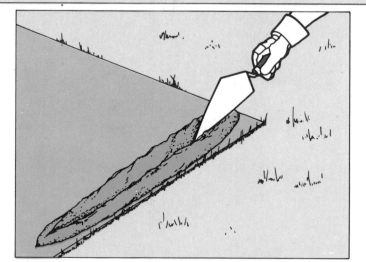

Laying the Brick

Beginning a row. Start at a corner, placing the first brick, dampened with a hose spray, on the mortar bed *(page 13)* just inside the end of the mortar line. Push the brick down about ½ inch into the mortar. Using a mason's level, check that the brick is aligned horizontally across its length and width, as well as vertically. If it is not, tap it with the trowel handle and test again, removing the brick and replacing it in the bed if necessary. Check with the level frequently while working.

Buttering intermediate bricks. Between the ends of a row, butter adjoining brick surfaces. For end-to-end bricks in a "stretcher course," scoop up enough mortar to cover the end; for side-to-side bricks in a "header course," scoop up enough to cover the side. Spread the mortar ¾ inch thick and remove any that slides over the edges.

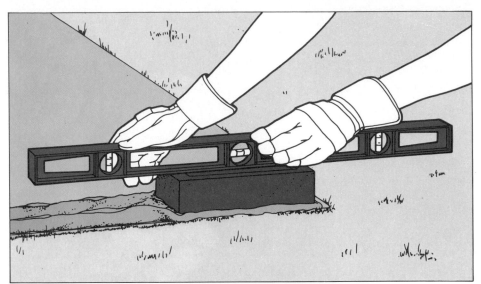

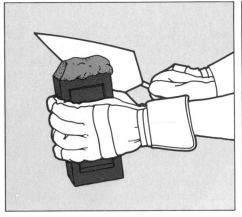

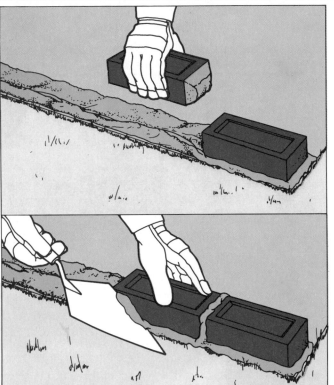

Laying intermediate bricks. Place the buttered brick in the mortar bed with the buttered end or side next to the end brick laid first *(upper drawing, right)*. In one motion, push the brick down ½ inch into the bed and toward the first brick to form a ½-inch joint between them. With the edge of your trowel trim off excess mortar that has oozed out of the joints *(lower drawing)*, and return this mortar to the mortarboard.

Laying the closure brick. Work from the ends of the course toward the middle, until there is room for only one more brick. Spread about a ¾-inch layer of mortar on the exposed ends of the two bricks last placed and, if necessary, add more mortar to the bed. Butter both ends of the closure brick, and gently insert it into the opening *(drawing)*. Push the brick into the mortar bed and trim off excess mortar at both horizontal and vertical joints. In an hour or two, finish the joints. Of the joint types sketched on the opposite page, each with the tool designed to shape it, only those that shed water well are recommended for general outdoor use. Use raked and struck joints only indoors or in dry climates.

Finishing the Joints

Concave. This is the most popular of all joints because it keeps moisture out and, since mortar is forced tightly between the bricks, makes an excellent bond. To shape it, press the mortar firmly with anything that fits: a convex jointer, a dowel, a metal rod or even a teaspoon.

V joint. This type stands out as a sharp line that directs water off. Form it with a V-shaped jointer, a bit of wood or the tip of a trowel.

Raked. Forming a deep recess, this joint is not waterproof but casts a dramatically heavy shadow to accentuate brick courses. To form the joint, take out ¼ inch of mortar with a raking tool and then smooth the recessed surface with a stick.

Weathered. This type sheds water more efficiently than any other joint shown here because it is recessed from bottom to top. It is formed by working from below the joint. Hold the blade of the trowel at a slant and compress the mortar from the front edge of the bottom brick upward to a point ¼ inch inside the top brick.

Struck. This joint is not water resistant because its recess slants from top to bottom. To shape it, hold the blade of the trowel at a slant and compress the mortar approximately ¼ inch away from the forward edge of the lower brick.

Flush. The easiest of all joints to make, this type is also water resistant. It is not particularly strong, however, because the mortar is not compacted. It is the joint that is left if you simply trowel off excess mortar after the brick is laid.

Extruded. Also called a weeping joint, this type is a particularly appropriate water-resistant finish for garden walls, or wherever a rustic appearance is desired. It is created by laying up new bricks with an excess of mortar; when the bricks are put into place the mortar is squeezed out and hangs down. To reproduce this effect when replacing old mortar, build up joints with excess mortar.

The Professional's Tricks for Simple Repairs

The monuments of the ancients prove that masonry can endure for millennia. But if its original attractiveness is to survive, cracks and crumbled sections must be repaired promptly—especially in northern climates. Even a crack 1/32 inch wide will allow moisture to penetrate a wall or slab and, by freezing and thawing, spread destruction. Fixing masonry is a simple process with mortar or grout. The techniques shown here for repairing brick, blocks, concrete and stucco apply to paths and patios as well as walls—only the direction of the work changes. But the method employed depends on the defect. Three types are most common: crumbling mortar in joints; a broken section of a wall; and deteriorating concrete steps, sidewalks and driveways.

If you notice crumbling mortar joints in walls, repair them by employing the process masons call pointing: chisel out the old mortar and replace it with new. If the bricks themselves are cracked or have broken out of the wall, replace both bricks and mortar. On both of these jobs you can make the repairs almost invisible by finishing new joints to match existing ones.

If new brick walls suddenly show long cracks running from top to bottom, the normal shrinkage of hardening mortar is probably to blame; such cracks can be cleaned out and pointed with mortar or filled with grout. If the cracks open up again after a few weeks or months, call a mason. Such recurrent cracking may indicate that the wall foundation is settling, and the wall could be pulled apart. Similarly, you can easily repair small cracks or broken patches in stucco walls but you should call in a professional if the en-

tire side of the house needs redoing.

Broken concrete sidewalks or cracked steps are too dangerous to leave unrepaired. The kind of material used to fix them varies with the dimensions as well as the type of defect (pages 20-21). Small cracks, for example, can be filled with grout; large ones require mortar. When large pieces, still intact, break off steps, you can paste them on again with a sand-cement-epoxy mortar. But when sections of steps crumble away, you will have to build them up again with conventional mortar (pages 22-23).

In any job remember that mortar and grout must be kept damp for a three- to four-day period to cure properly and form a strong bond. Keep the repair covered if possible, and sprinkle mortar or grout with water as necessary—even four or five times a day in hot, dry weather.

Replacing Damaged Mortar

1 Cleaning out the joint. With a cold chisel and ball-peen hammer or maul, remove crumbling mortar from a joint to a depth of at least 1 inch. Then chop out enough additional old mortar to expose bare brick on one side of every joint. Brush out the joints, or clean them by blowing sharply into them. Caution: goggles are absolutely necessary to protect your eyes.

2 Laying the mortar. Dampen the joints slightly with a small, wet brush or a garden hose set at a fine spray. Spread a ½-inch-thick layer of mortar (page 12) onto a hawk, a small mortarboard that has a handle so it can be easily held up to the work. With the bottom edge of a pointing-trowel, which is smaller than a mason's ordinary trowel, slice off a thin piece of mortar; lift it

from the hawk and press it into the joint. Because you push mortar away from you when pointing, always use the bottom side of the trowel. For extra-long horizontal joints—many bricks wide—you may want to use a joint filler.

Splitting Brick

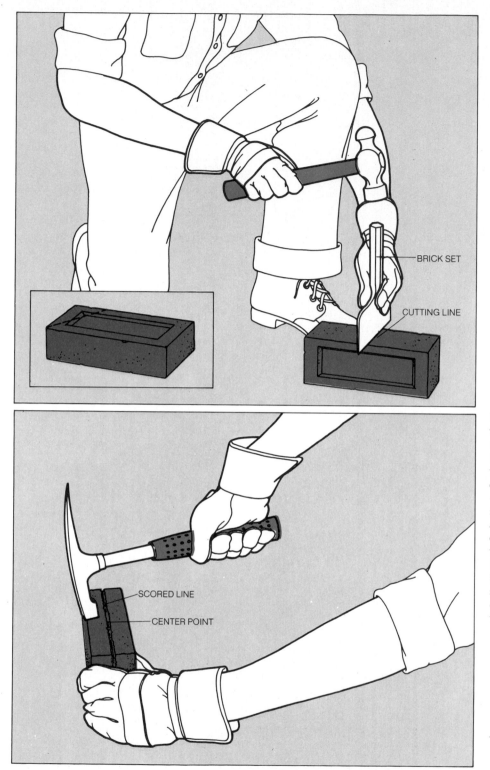

BRICK SET

CUTTING LINE

SCORED LINE

CENTER POINT

Crosswise cuts. Most brick jobs will probably require you to cut some bricks into smaller sections. It is easiest to cut them all at once. With a pencil and ruler, mark a cutting line across the long side edges of each brick. Mark on the diagonal (*inset*) if this shape is required. Put on goggles to protect your eyes from chips, and score along the cutting lines with a brickset, a broad-bladed chisel. Place the blade end of the brickset—beveled edge facing away from the part of the brick to be used—on the cutting line. Then tap the end of the handle with a ball-peen hammer.

Before cutting the brick, lay it on a bed of sand to cushion it. The long side edge should face up, and the part of the brick to be used should point toward you. Insert the brickset into the uppermost scored line, with the bevel again facing away from the part to be used. Then strike the handle with a sharp blow of the mallet. The brick will separate into two pieces.

Lengthwise cuts. Find the center point of one of the long side edges of the brick. Measure and mark a continuous cutting line all around the brick. Wearing goggles, score the surface along the cutting line by tapping the line with the sharp edge of the square end of a bricklayers' hammer —or, for extra-hard bricks, with a brickset—as shown above. Grip the brick firmly in one hand, and strike the brick sharply with the flat side of the square end of the hammer, just beyond the center point of the scored line. (You may have to practice this technique several times before you will be able to halve the brick with a single blow.) Use the curved, chisel-like blade end of the hammer to clean off rough spots on the cut edges, removing small bits at a time.

Replacing a Broken Brick

1 Removing a single damaged brick. Rip out mortar with a mortar rake—a valuable tool if you can find one—or a cold chisel and ball-peen hammer. Because the mortar rake enables you to use both hands, it speeds up the job. Chisel out the damaged brick and brush the space clean.

2 Replacing the brick. Select a brick that fits the slot or cut one to fit *(page 17)*. Dampen the slot's surfaces and apply a thick coating of mortar. Hold the brick on a hawk about ½ inch above the course the brick will rest on. Ram the brick into the slot. Trowel in extra mortar if needed to fill the joints.

Filling in a Damaged Wall

Cutting out and replacing bricks. Remove all mortar surrounding damaged bricks and chop out all broken or cracked bricks with a broad chisel (also called a brickset) and a ball-peen hammer. When removing adjoining bricks, work from the top of the damaged area down. Brush away bits of mortar, brick and dust. Then, working from the bottom course upward, dampen all surfaces of both old and replacement bricks, and lay mortar beds for the replacement bricks, troweling and furrowing the bed just as you would for placing new bricks. Butter the bricks and lay them in place on the mortar beds. When the mortar is firm, finish the joints to match the rest of the wall *(page 15)*. Keep the new mortar wet for three or four days to cure it.

Patching a Large Vertical Crack with Grout

1 Pouring the grout. A long, broad, vertical crack, such as often appears in an exterior chimney wall, can be repaired quicker by pouring it full of grout than by laying in mortar with a trowel. First clean the crack. Mix brick mortar in a bucket, adding enough water so that the mortar flows easily. Dampen the bricks around the crack with a wet brush. Then cover the lowest section of the crack with a piece of wide heavy-duty adhesive tape, 12 to 18 inches long. Pour grout from the bucket into the taped section, using a funnel *(below)*. Work upward section by section to fill more of the crack in the same way, but do not try to repair more than about 3 feet at a time.

2 Shoring up the grout. A crack repaired with tape-held grout needs reinforcement. Brace a board over the tape until the grout sets, usually two to three hours. Provide the reinforcement right after filling each section if the tape does not adhere well—on rough bricks for example—otherwise the board backing will not be required until you have filled about 3 feet of the crack.

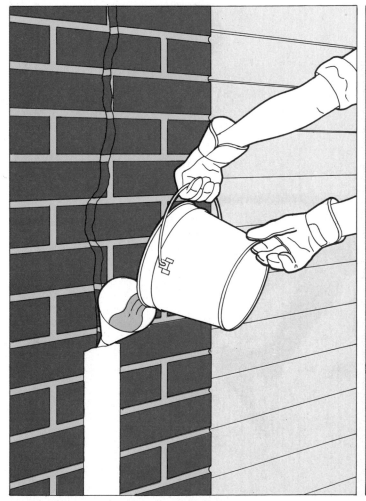

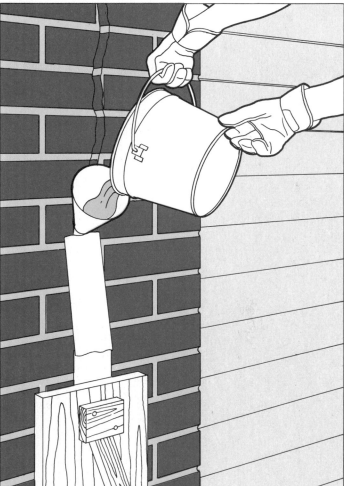

How to Restore Concrete

Paradoxically, concrete surfaces cannot be repaired with concrete—the coarse gravel aggregate in the new concrete would prevent a strong bond between the patch and the damaged area. Instead, the materials used are grout, mortar, sand-cement-epoxy compounds, or pre-mixed cement or latex compounds that come in a can or cartridge.

Before using any of these materials, clear the repair area of damaged concrete, keep the area moist for several hours—preferably overnight—and remove all dirt, debris and standing water. Then apply a patch of the appropriate material. For cracks up to ⅛ inch wide, use either a grout made of portland cement and just enough water to make a paste that will hold its shape, or use a ready-mix. Cement-based ready-mix in a cartridge can be applied directly with a caulking gun; force a canned mix or grout into the crack with a putty knife or a mason's trowel, and smooth it flush to the level of the surrounding concrete.

Mend larger cracks with mortar consisting of 1 part portland cement, 3 parts sand, and water; a latex-based ready-mix; or a sand-cement-epoxy mix. The epoxy compound produces an especially durable repair on spalled surfaces from which concrete has flaked off in thin scales. Wear gloves when repairing concrete, and goggles when breaking up the damaged surface.

Epoxy mixtures cure by themselves in 24 hours, but a grout or mortar patch must cure slowly in the presence of moisture. On a horizontal surface let the patch set for about two hours, then cover it with a sheet of plastic secured by bricks or rocks. For the next five to seven days, lift the cover daily and sprinkle water on the patch. If a patch on a vertical surface cannot be covered conveniently, sprinkle it twice a day.

Treating Wide Cracks

1 **Removing the damaged concrete.** Chip away all cracked or crumbling concrete to about 1 inch below the surface, using a cold chisel and a ball-peen hammer. Wear goggles during this and the following step to protect your eyes from flying fragments of concrete.

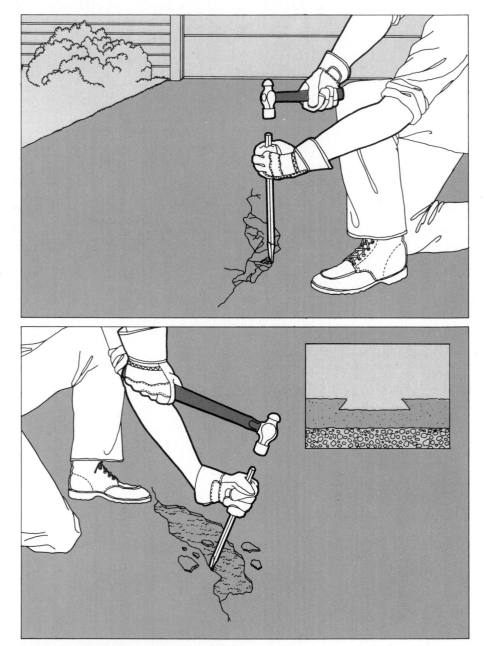

2 **Undercutting the edges.** To provide a better bond and keep the patch from heaving upward after the job is done, undercut the edges of the crack with the hammer and chisel until the hole you have made is wider at the bottom than at the top (*inset*). Remove all rubble and dirt. Soak the crack with water for several hours; if possible, run a trickle from a garden hose through it overnight.

3 **Coating the edges with cement paint.** Thoroughly mix 1 part portland cement and 3 parts sand; add sufficient water to make a paste stiff enough to work with a trowel, and set it aside. Then mix a small batch of portland cement and water to the consistency of thick paint. Coat the edges of the crack with this cement paint and proceed immediately to Step 4. Caution: You must complete the repair before the paint dries.

4 **Filling the crack with mortar.** Pack the mortar firmly into the crack with a pointed mason's trowel, cutting deep into the mixture to remove air pockets. Level the mortar with a square trowel, let it stand for an hour, then spread it evenly back and forth across the crack, always with the leading edge of the trowel slightly raised *(drawing)*. To cure the patch, follow the procedure described in the text on the opposite page.

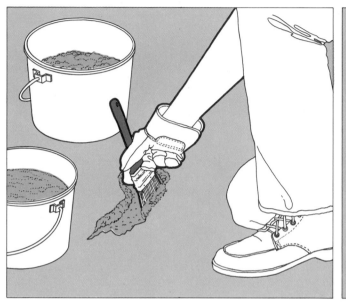

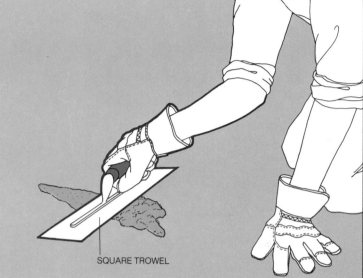

SQUARE TROWEL

Refinishing Spalled Surfaces

Using an epoxy mix. Wearing goggles to protect your eyes from flying fragments, break up large areas of scaling concrete with an 8-pound sledge-hammer (let the hammer drop of its own weight, rather than swinging it hard against the surface); for small areas, use a ball-peen hammer and a cold chisel. Sweep up dust and debris, using a stiff wire brush, if necessary, to dislodge any small fragments. Wet down the damaged area and keep it moist for several hours, preferably overnight. The area should still be damp when you begin to apply the patch.

Prepare a sand-cement-epoxy mixture according to the manufacturer's directions. Apply the mixture with a square trowel, bringing the new layer level with the surrounding concrete, and feathering it thinly at the edges. Let the patch stand for 24 hours before putting pressure on it.

Putting Back
a Broken Step

1 Gluing the broken piece. Brush particles of dirt and cement from the broken piece and the damaged corner of the step. Mix a small batch of a sand-cement-epoxy glue, following the label directions and, with a mason's trowel, coat the part of the piece that will face the step. Position the piece and hold it firmly for about 10 minutes, until the glue has hardened; if necessary, prop a board against the piece to hold it in place.

2 Completing the job. After the glue has set, use a scraper or putty knife to remove any excess that has oozed out between the piece and the step. You will probably find a small, irregular crack around the repair; patch it as you would any narrow crack, using the epoxy glue instead of grout. Avoid stepping on the repaired corner or bumping against it for at least 24 hours to allow the glue to harden completely.

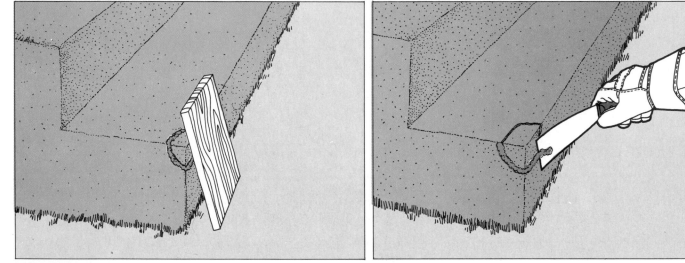

Rebuilding the
Corner of a Step

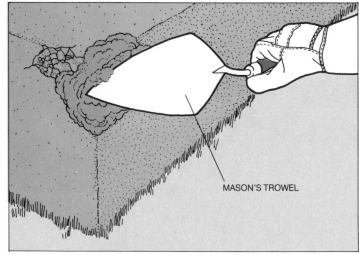

MASON'S TROWEL

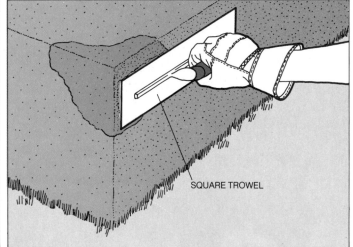

SQUARE TROWEL

1 Shaping a replacement piece. If the original corner of the step has crumbled away or been lost, clean and moisten the corner. Thoroughly mix 1 part portland cement with 3 parts sand and add just enough water to make a paste that holds its shape. With a mason's trowel, apply the mortar to the step in the rough shape of the original corner. Let the mortar harden until it retains a firmly impressed thumbprint (this may take up to five or six hours, depending on weather conditions).

2 Finishing the corner. Finish and smooth the corner flush to the adjoining parts of the step with a square trowel. Let the mortar cure for a week, moistening it twice a day, and avoid stepping on the corner or bumping it for another three weeks.

Repairing the Chipped Edge of a Step

1 Clearing the damage. With a ball-peen hammer and a cold chisel held horizontally, chip off the damaged concrete all the way across the edge of the step. Be sure to wear goggles to protect your eyes from flying fragments.

2 Undercutting a groove. Still wearing the goggles, but holding the chisel at an angle, chip away enough of the step to make a V-shaped groove *(inset)*. Clean off all debris and keep the edge damp for several hours, preferably overnight.

SQUARE TROWEL

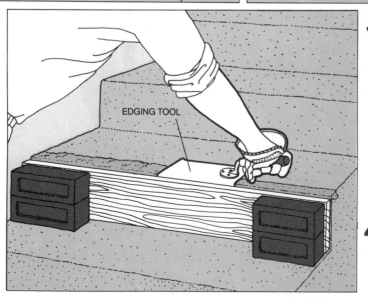

EDGING TOOL

3 Rebuilding the edge. Thoroughly mix 1 part portland cement with 3 parts sand, then add just enough water to make a paste that holds its shape; set this mortar aside. Make a form board as wide and high as the riser and set it against the step, holding it in place with heavy objects such as bricks or concrete blocks. Coat the edge of the step with cement paint *(page 21, Step 3)*, then immediately fill the V-shaped groove with mortar, shaping and smoothing it flush to the adjoining surfaces with a square trowel.

4 Completing the job. Let the mortar set for about an hour, finish the step to a rounded edge with an edging tool, then carefully remove the board. Keep the area damp for at least a week to help the mortar cure, and do not step on the new edge for another five to seven days.

Renovating Damaged Stucco Walls

Because stucco is a form of concrete used for exterior coatings, it can often be repaired by much the same method as ordinary concrete. A damaged area larger than 1 or 2 square feet usually cannot be repaired; the entire wall must be resurfaced, a job for professionals. But small patches can be made that will adhere to any stucco base—metal lath, ordinary concrete, brick or stone.

The mix for such patches consists of 1 part portland cement, 3 parts sand and ¹⁄₁₀ part hydrated lime, measured by volume, with just enough water added to give the consistency of putty. On a painted wall, follow the patching procedure on this page, then paint the patch to match the surrounding area. If the original stucco is not painted but is colored with embedded pigments, use white cement and white sand instead of the portland cement and ordinary sand of the finishing mortar and add metallic oxide pigments up to 2 per cent of the total mix by volume. Matching the original color by this method is difficult. Mix a small batch of mortar and let it dry before judging the color match, then make closer matches by trial and error. Or, if your supplier stocks it, make an approximate match with stucco finish—a pigmented dry mix that comes in a range of pastel colors.

You can give a patch a textured finish with a wooden float (Step 3, right). To create a finer texture, fasten a piece of carpet or plastic foam to the base of the float; for a coarser finish, score the surface with a whisk broom. If the surrounding stucco has a rough, pebbled surface, embed colored marble chips or gravel in the patch before it hardens.

1 Preparing the surface. Wearing goggles to protect your eyes from flying fragments, use a putty knife to scrape away the damaged stucco until you reach a firm layer—if necessary, remove stucco down to the base (in the drawing, the base is metal lath). Undercut the edges of the repair area, remove loose material with a wire brush and keep the area damp overnight.

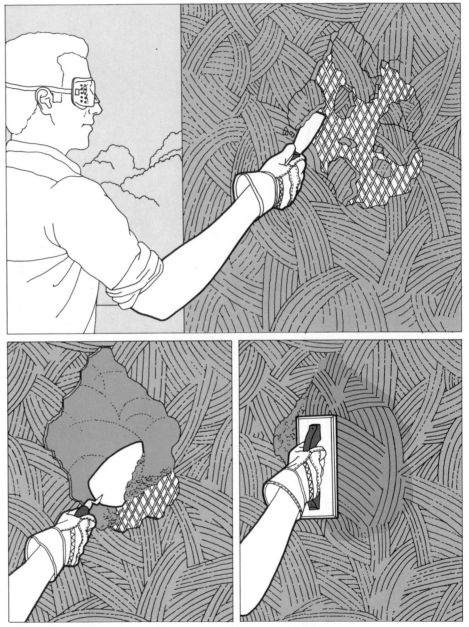

2 Building the patch. Mix a batch of mortar to match the surrounding wall. Trowel the mortar into the damaged area using the back of the trowel to pack it in as firmly as possible. If you have had to expose the base, apply the mortar in two layers. First pack the area with mortar to within ¼ inch of the surface and let it set for two days; then moisten the area again and apply a finishing coat of the same mortar, leveled with a square trowel.

3 Applying a texture. To give the patch a coarse, sandy texture, rub a wooden float—the tool ordinarily used to finish a masonry surface—over the surface in small, circular strokes. Keep the patch damp for three days, sprinkling it twice a day with the fine spray of a garden hose. If the wall is exposed to strong sunlight or heavy winds, hang a piece of wet burlap over the patch and keep it damp to slow the curing process.

First Aid for Holes in Asphalt

The asphalt used to blacktop driveways and walks is actually a kind of concrete, a mixture of gravel with a binder. However, the binder is different—a crude-oil extract rather than cement holds the gravel together, giving asphalt its characteristic black color. Like any concrete, asphalt can develop cracks and holes from frost or ice-melting salts; it is vulnerable also to such substances as oil dripping from a car. To protect asphalt against most forms of deterioration, coat it once every four or five years with a waterproof and chemical-resistant sealer. Simply pour sealer straight from the can onto the driveway and spread it evenly, like thick paint, with a push broom.

The coating process will also fill any cracks up to ⅛ inch wide that have appeared in the asphalt. For larger cracks —up to ½ inch wide—pour ready-to-use crack filler into the cavities one week before spreading the sealer. And for cracks up to an inch across, thicken sealer with sand to a putty-like consistency, push the mixture into the crack with a putty knife, then seal over it.

While sealer will fill most cracks, holes in a blacktop surface must be filled with "cold-mix" asphalt as illustrated at right. Cold-mix comes in two varieties: cutback asphalt and emulsion mix. Either type is satisfactory for dry holes, but damp holes require emulsion mix, which is made with water. Both types come ready-mixed in 66-pound bags, enough to patch about 1⅓ square feet of asphalt to a depth of 4 inches.

If the temperature is below 40°, asphalt should not be patched or sealed. If cool weather has hardened cold-mix into an unworkable lump, soften it by standing the bag next to a water heater for a couple of hours before opening it.

How to Use Prepared Cold Mixes

1 Preparing the hole for repair. With a shovel, dig out the hole to a depth of 3 or 4 inches and remove any loose material. Cut back the edges of the hole to sound asphalt, making sure that the sides of the hole are vertical. Compact the bottom of the hole with a tamper made by fastening a pair of large door handles to opposite sides of a 4- or 5-foot-long 4-by-4.

2 Adding the asphalt. Fill the hole halfway with cold-mix asphalt. Slice through the asphalt with the shovel to open air pockets, then compact the asphalt with the 4-by-4 tamper (bottom left). Add ½-inch layers of cold-mix, tamping each time, until the top of the patch is even with the surrounding blacktop. (If car wheels will run over the patch, make it ½ inch higher than the rest of the driveway; the weight of the car will flatten the hump.) Spread sand over the new asphalt until it dries (usually about two or three days) to keep it from sticking to your shoes.

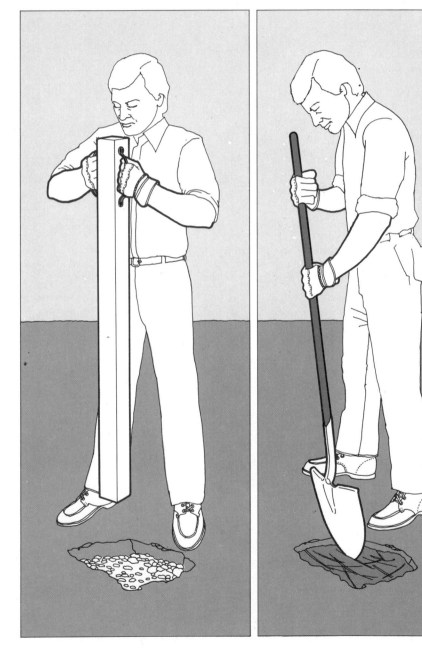

Homemade Concrete in Convenient Batches

Concrete is man-made stone, and like stone it can last forever. Concrete steps or a post anchored in concrete will stand up to almost any kind of punishment. The great strength comes from the materials: gravel (called coarse aggregate), sand and cement. The coarse aggregate supplies bulk. The sand fills voids between the coarse pieces. And the cement, when it is moistened with water, binds the sand and aggregate into a durable monolithic solid.

For anchors on posts, clothesline supports and backyard gym poles, you can make concrete of plain portland cement. But for larger jobs such as building concrete patios or steps, you need air-entrained portland cement, which contains additives that produce and trap, or entrain, microscopic air bubbles in the concrete. When the concrete becomes wet, the bubbles act like tiny ball bearings to facilitate pouring and spreading. When the concrete dries, the bubbles form tiny spaces within the concrete so that it can expand and contract with a minimum of cracking. Because of this weather resistance, some local building codes require air-entrained cement for concrete that will be outdoors.

Cement comes in one-cubic-foot bags that weigh 94 pounds each (80-pound Canadian bags contain ⅞ cubic foot). Be sure the cement is dry before you use it. If a bag is hard around the edges, its con-tents have probably absorbed moisture but are acceptable—rolling the bag on the ground will generally break up the lumps. If the edges do not break this way, however, or if you find any lumps inside that do not break between your fingers, return the bag to your dealer.

Sand and coarse aggregate are also sold by the cubic foot. Sand for concrete, unlike the uniform sand needed for mortar, should contain particles of all sizes from dust to about ¹⁄₁₆ inch in diameter. Do not use beach sand; the salt in it will weaken concrete. Gravel aggregate should contain particles no larger than 1 inch in diameter for thin layers, such as you may need for a small fishpond; use 1½-inch aggregate where sections more than 4 inches thick are required, in such structures as steps, patios and sidewalks.

The amount of water required for a concrete recipe is critical; even a small amount of extra water can weaken the concrete. Most recipes are based on what is called wet sand, which forms a ball when squeezed in the hand but leaves no noticeable moisture on the palm. Since the sand you buy rarely matches this moisture level, you must adjust the amount of water added during mixing (page 28). The three recipes given below will produce variations of a basic 1:2:4 concrete mix—one part cement, two parts sand and four parts coarse aggregate—that is strong and workable yet viscous enough to be poured. To stiffen the mix for spreading rather than pouring —for making a bowl-like fishpond, for example—use about ⅔ the amount of water, added gradually only until the ingredients are thoroughly combined.

How you mix the materials depends on how much concrete you need. A small amount of plain (not the air-entrained type) concrete can be mixed by hand in a wheelbarrow, or on a driveway or other flat surface. But to mix enough air-entrained concrete to build steps or a footing for a wall, rent a gasoline- or electric-powered mixer and set it up close to your work area. Working with a power mixer, you probably will be able to pour and finish about 10 square feet of a six-inch thick slab before the concrete hardens too much to be finished. If you require larger quantities to be available, arrange for ready-mixed concrete delivered in a truck.

When mixing by hand, use a container marked in gallons to measure out and transfer the materials from bag to wheelbarrow or driveway work space (chart, below). For larger jobs requiring a power mixer, you may want to make a bottomless box, 12 inches square and 12 inches deep, to measure one cubic foot of dry ingredients. Set the box in a wheelbarrow, shovel the box full and lift the box out; then shovel the cubic foot of material into the mixer.

Three Basic Recipes

Ingredients to make one cubic foot of plain concrete:	Ingredients for one cubic foot of air-entrained concrete, using ¾-inch coarse aggregate:	Ingredients for one cubic foot of air-entrained concrete, using 1½-inch coarse aggregate:
1⅝ U.S. gallons plain portland cement (20 pounds or ⅕ cubic foot)	1⅝ U.S. gallons type 1A portland cement (20 pounds or ⅕ cubic foot)	1⅜ U.S. gallons type 1A portland cement (17 pounds or ⅙ cubic foot)
3¼ U.S. gallons sand (40 pounds or ⅖ cubic foot)	2⅞ U.S. gallons sand (35 pounds or ⅜ cubic foot)	2⅞ U.S. gallons sand (35 pounds or ⅜ cubic foot)
6 U.S. gallons ¾-inch gravel aggregate (80 pounds or ⅘ cubic foot)	6 U.S. gallons coarse aggregate (80 pounds or ⅘ cubic foot)	6⅜ U.S. gallons coarse aggregate (85 pounds or ⅞ cubic foot)
1¼ U.S. gallons water (10 pounds)	1¼ U.S. gallons water (10 pounds)	1¼ U.S. gallons water (10 pounds)

Getting the Materials Ready to Use

Keeping the ingredients separate. Deposit cement, sand and coarse aggregate close to the work site. To prevent cement from absorbing moisture, stack bags against one another on a raised platform away from walls. On plastic sheeting, or a tarpaulin, pile sand and aggregate into several small adjoining mounds. Keep them apart with a board divider if necessary. If the materials must be stored longer than a few hours, place both open and unopened cement bags inside plastic bags, stack the bags on the platform and cover everything with a waterproof tarpaulin.

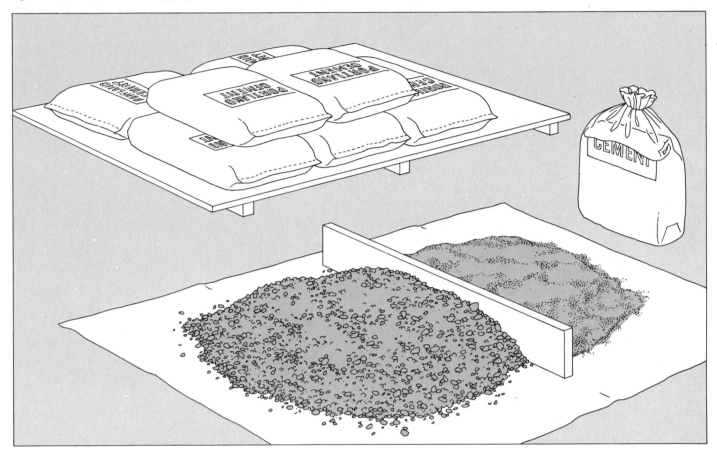

An Extra Ingredient: Built-in Color

Concrete does not have to be plain gray. You can coat the top with a dusting of pigment or a layer of colored mortar while it is still wet *(pages 50-53)*. And you can, of course, paint or stain it after it cures. But the best way is to build the color in by using pigmented cement when you mix the concrete.

Using pigmented cement distributes particles of color evenly throughout. This method, though, requires the use of a power mixer. If you try to mix pigmented cement with the aggregate and water by hand, the result is likely to be a mottled, not uniform, shade.

Precolored cement may be hard to find (and expensive) but you can create your own by adding synthetic or natural pigments to either gray or—for clearer tones—white cement. The synthetic pigments cost more than the natural metallic oxides, but are less likely to bleach or fade. Both types come in a range of colors you can use as is, or combine to produce custom shades.

To keep coloring uniform from batch to batch, measure pigments—and all the other concrete ingredients—by weight, not volume. Use a bathroom scale, wrapped in clear plastic to keep it clean and readable, for the weighing; three- to five-gallon containers are a convenient size to hold the ingredients.

Concrete lightens as it dries, so finding the right amount of pigment to use will take experimentation. In general, about one pound of pigment and one bag of white—not gray—portland cement will produce a pastel; one to three pounds to one bag of cement will produce a medium shade; four to nine pounds to one bag of cement will produce a dark color. To avoid weakening the concrete, never add more than 10 per cent pigment.

Mixing Concrete in a Wheelbarrow

1 Measuring the materials. Make no more than one wheelbarrow-sized batch of concrete—usually 2½ cubic feet—at a time. Measure out each of the ingredients (*recipes, page 26*) in a bucket (one marked in quarts is most convenient).

2 Combining the dry materials. Pour the sand into the wheelbarrow, and, using a hoe, shape the sand into a ring. Pour the cement into the center of the sand ring, then mix the ingredients. When the mixture is uniform in color—without streaks of brown or gray—shape another ring. Pour coarse aggregate into the center. Mix until the coarse aggregate is evenly distributed.

3 Adding the water. Push the sand-cement-coarse aggregate mixture to the sides of the wheelbarrow to form a bowl-like depression. Slowly pour about three quarters of the water into the depression. Pull the dry materials from the edges of the ring into the water, working all around the pile until the water is absorbed by the mixture. When no water remains standing on the surface, turn the concrete three or four times. Add the remaining water a little at a time until the mixture completely coats all the coarse aggregate. Leave any unused water in the measuring bucket until after you test the consistency of the concrete and make necessary corrections (*right*). At the end of the day, clean up as explained on page 37.

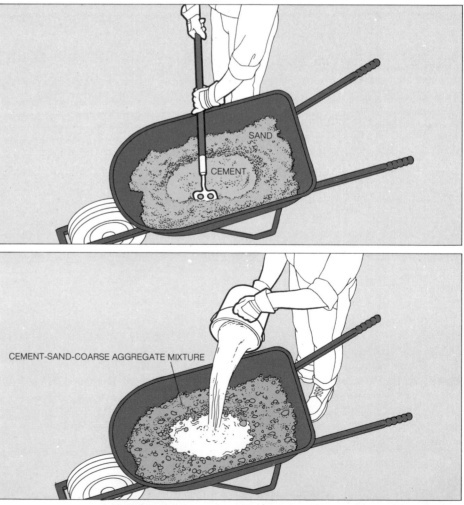

A Power Mixer to Make the Job Go Faster

Before turning on the mixer, set the drum to the mixing position; then pour in half the water *(recipe, page 26)* and all the coarse aggregate. Turn on the mixer and add—in order—sand, cement and about half the remaining water. Wait three minutes, then add more water, little by little, until the mixture completely coats the coarse aggregate and the concrete is a uniform color. Test a few shovelfuls in a wheelbarrow *(below),* and return the test batch to the mixer before making corrections. When the concrete is properly compounded, dump the mixerful into your wheelbarrow and hose out the drum. When you finish using the mixer, clean up as described on page 37.

Testing the Consistency

Judging and correcting the mix. Smooth the concrete in your wheelbarrow by sliding the bottom of a shovel or trowel across the concrete's surface. Then jab the edge of the shovel or trowel into the smooth surface to form grooves. If the surface is smooth and the grooves are distinct, the concrete is ready to use *(left, bottom).* If the surface roughens or the grooves are indistinct, add no more than a half-gallon of a cement-water mixture, using twice as much cement as water. If the surface is wet or the grooves collapse, add no more than a half-gallon of the cement-sand mixture combined in the original proportions. Retest the batch until the consistency is correct.

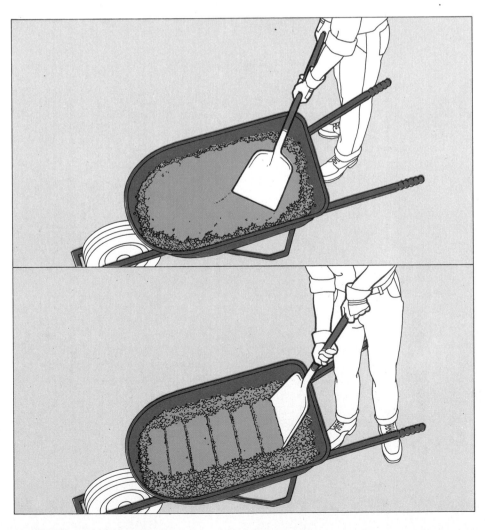

Solid Bases for Posts and Poles

The strongest way to set a post in the ground is to anchor it in concrete. Thus secured, fences will stand straight longer, clothesline supports and flagpoles will resist toppling, and backyard playground equipment will become tipproof so that it is safer and more fun for children to play on. For the monkey bar type of yard toy, a concrete anchorage is by far the safest installation.

Every post-setting job begins with a hole. A manual post-hole digger, made of two shovels hinged together like pincers *(right)*, can produce holes 6 inches in diameter or larger and up to 3½ feet deep. If you must dig more than half a dozen or so holes, you may want to rent a gasoline-powered auger.

Manufacturers of fencing and other outdoor equipment usually suggest how big a hole to dig. The hole should be deep enough to encase one third the length of the post (but no more than 3 feet). The width of the hole should allow enough clearance all around the post for pouring concrete. If the hole does not extend below the frost line, taper the hole to widen the bottom a few inches to a bell shape *(bottom right)* so that frost is less likely to heave the post.

To keep wood posts from rotting, soak them overnight in one or another of the commercially available standard preservatives. Coat the buried ends of aluminum and steel with bituminous paint to prevent corrosion. Then simply pour concrete around the posts—except in the case of flagpoles and clothesline supports. They usually come with a metal sleeve that is anchored for the pole itself to slip into (and out of). For a long two- or three-section flagpole that would be awkward to lift from its sleeve, a third anchoring alternative is a hinge support *(overleaf)*.

For any of these anchoring methods, use the recipe for hand-mix, plain concrete on page 26. How much concrete to mix depends on the size of the hole. A 6-inch hole holds $^2/_{10}$ cubic foot of concrete for each foot of its depth, an 8-inch hole takes $^4/_{10}$ cubic foot for each foot of depth and a 12-inch hole, $^8/_{10}$ cubic foot for each foot of depth.

Setting Fence Posts

1 Digging the holes. Tie a string guide between stakes set about a foot outside the ends and corners. With a post-hole digger or a power auger, dig holes at the intervals specified for your fencing. Place 6 inches of gravel in the bottom of each hole for drainage.

2 Installing a post. To align a wood post, set two stakes next to the hole and fasten a 1-inch board brace to each stake with a single nail. Set the post in the hole, get it vertical with a level and nail the braces to the post. Pour concrete around the post, overfilling the hole slightly. To install a metal post, have a helper hold it centered in the hole while you pour the concrete. Then hold a level alongside the post to align it. When standing water has evaporated from the concrete, bevel the surface downward with a trowel so rain water will run away from the post. Wait a day before removing the braces or attaching fencing. If the concrete pulls away from the post, pour grout into the gap.

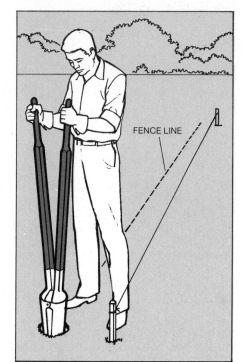

FENCE LINE

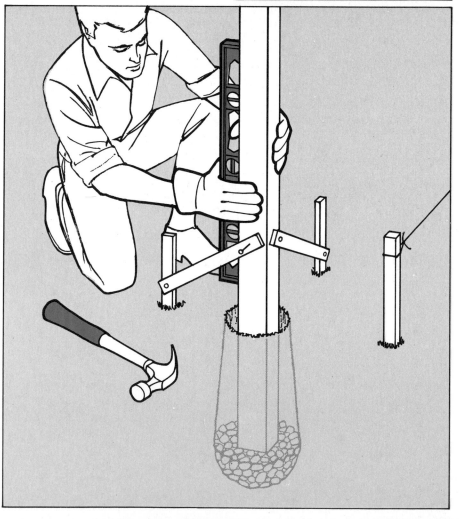

Anchoring a Sleeve

1 Forming a collar. To make a neat edge for a flag-pole anchorage, allow for a collar to hold the concrete in place. Dig a hole about 4 inches deeper than the length of the sleeve. Spread a 6-inch layer of gravel in the bottom of the hole so that the sleeve will extend about 2 inches above ground when installed. Shape a strip of 6-inch-wide lawn edging into a collar slightly larger than the rim of the hole. Push it into the hole until the top edge is even with the lip; the metal's springiness will hold the collar in place.

2 Pouring the concrete. Hold the sleeve centered in the hole and pour the concrete, overfilling the hole slightly. If the sleeve shifts during pouring, recenter it and align it vertically before the concrete sets. You can generally judge by eye if a sleeve for a short pole is vertical, but tall poles may require precise alignment as described at bottom. Finish concrete as described for a fence post *(opposite, bottom).*

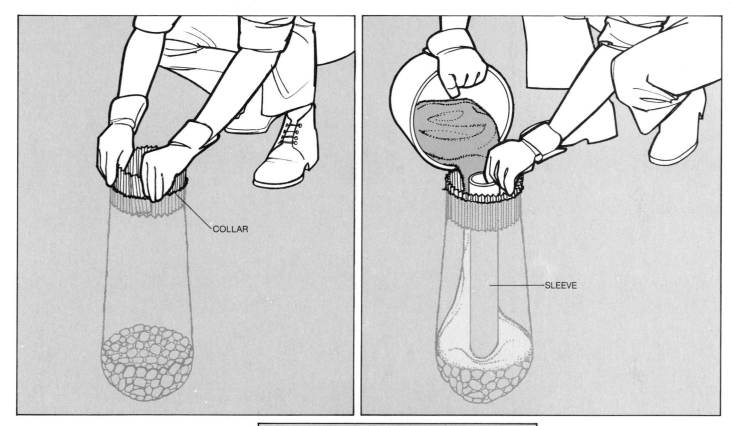

COLLAR

SLEEVE

3 Making a sleeve precisely plumb. Since even a small tilt may be noticeable in a long pole, adjust alignment of the concrete-encased sleeve with the help of a plumb bob. Hang the bob inside the sleeve from a stick laid across the sleeve top, then adjust the sleeve so that the bob is centered over the bottom. After the pole has been installed, minor adjustments can be made by hammering wood wedges into the gap between the sleeve and the side of the pole.

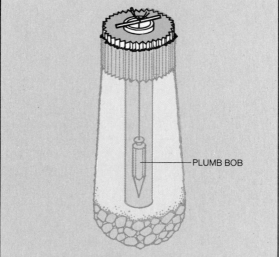

PLUMB BOB

A Hinge for a Tall Pole

1 Making the hinge. Raising and lowering a tall, multisection pole is easier if it pivots on a pair of steel supports made from two lengths of 3-inch-wide and ³⁄₁₆-inch-thick channel iron, each ⅕ the total length of the pole. A welding shop can supply the channel, and can drill ¾-inch holes in the positions indicated in the drawing. Take along the pole and have it drilled at the same time. Assemble the hinge by bolting the channel irons and bottom pole section together with ⅝-inch galvanized bolts as shown at right.

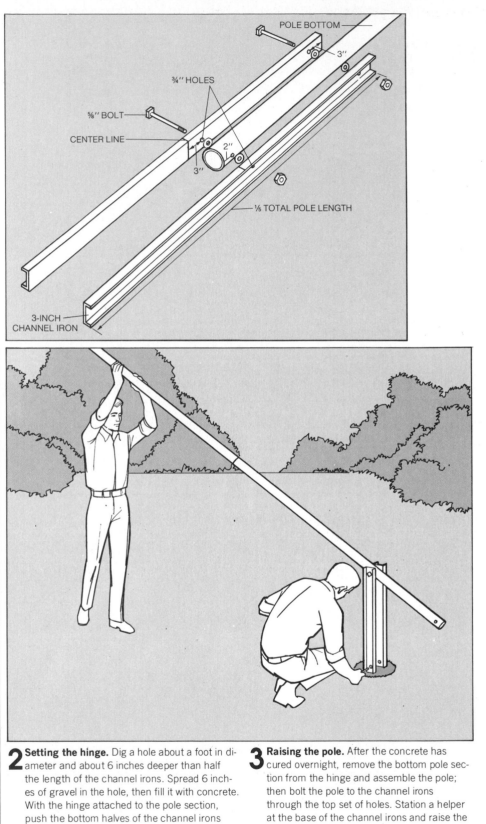

POLE BOTTOM

3″

¾″ HOLES

⅝″ BOLT

CENTER LINE

2″

3″

⅕ TOTAL POLE LENGTH

3-INCH CHANNEL IRON

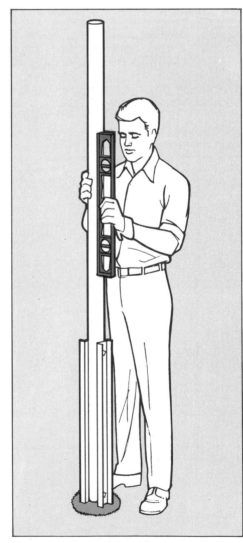

2 Setting the hinge. Dig a hole about a foot in diameter and about 6 inches deeper than half the length of the channel irons. Spread 6 inches of gravel in the hole, then fill it with concrete. With the hinge attached to the pole section, push the bottom halves of the channel irons into the concrete, holding a level against the pole to align it. Bevel the surface of the concrete downward. Leave the pole section bolted to the hinge while the concrete sets.

3 Raising the pole. After the concrete has cured overnight, remove the bottom pole section from the hinge and assemble the pole; then bolt the pole to the channel irons through the top set of holes. Station a helper at the base of the channel irons and raise the pole so that he can insert the second bolt.

Tipproofing a Backyard Gym

1 **Preparing the holes.** Assemble the gym and drive a stake into the ground at each leg. Move the gym aside, then dig a hole centered on each stake, 10 inches in diameter and 1½ feet deep.

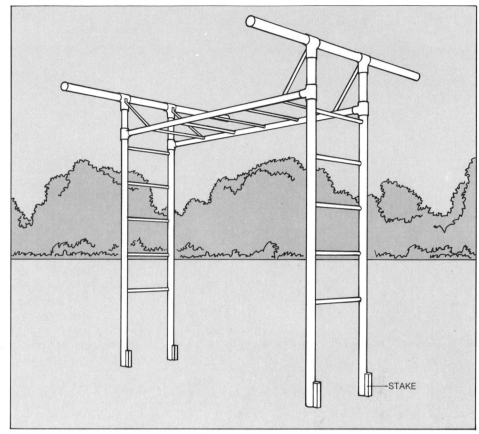

2 **Pouring the concrete.** Spread 6 inches of gravel in the bottom of each hole, then set the gym into the holes. Level the gym by adding or subtracting gravel, and fill the holes with concrete. Bevel the top of the concrete away from the legs. Caution: Do not permit the gym to be used until the concrete has cured for a couple of days.

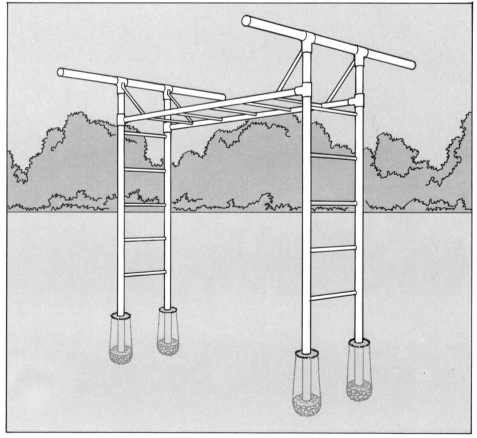

Making Holes and Securing Fasteners in Masonry

Many home improvement projects, such as putting up a porch railing *(below)*, installing outdoor wiring or running a water pipe to an outside faucet, require holes in masonry, special supports or both. The holes are easy to make, but you must wear goggles. Various fasteners are designed for masonry—screw anchors, toggles, adhesive fillers and even nails shot out of a pistol.

Most fasteners fit into predrilled holes. Small holes in cinder block and plaster can be made with a carbide-tipped masonry bit in an ordinary electric drill. For holes larger than an inch, use a four-edged chisel called a star drill, or simply knock through into a hollow space with a hammer and patch up the irregularities later with mortar.

For 1-inch or smaller holes in concrete, brick or stone, rent an electric hammer-drill from a tool-rental agency and use it with a carbide-tipped bit. Hammer drills are constructed so that they pound the bit into the masonry about 3,000 or more times per minute, giving the tip a sharp bite. Holes larger than an inch are most easily made by using a masonry-core bit—similar to a hole saw for wood—in a hammer drill *(opposite)*.

Which fastener you use depends partly on the object you are mounting, partly on the wall. Screws and expansion anchors, toggle bolts, steel masonry nails, and a family of glues called mastics are the principal types.

For most purposes, expansion anchors and toggles of the kind used in interior plaster and plasterboard serve in masonry. Screws driven into expansion anchors are suitable for concrete and the solid parts of blocks. They can be used in mortar joints and in brick if you are careful not to tighten the screw so much that the material around the edge of the hole begins to crumble. Toggles are needed for the hollow parts of blocks.

In stone, expansion anchors are unsatisfactory; they can create stresses that will cause cracks. For this material—and any solid masonry where a very strong fastener is required—a technique employed in highway construction to anchor concrete slabs is useful. A stud is made from an ordinary galvanized bolt by securing the head in a hole with epoxy, and the object is secured to the bolt with a nut.

Nails can be driven by hand or with a special tool *(page 36)*. Mastic glues are simply spread on the masonry surface. Both are best reserved for attaching light loads, such as furring strips for paneling, to block and concrete. Brick and stone, however, are too hard for nailing except at mortar joints; they also make relatively poor surfaces for gluing.

Locating holes in masonry. Accurate marking for holes in masonry is especially important since drilling errors are virtually impossible to correct. Carefully check measurements and, wherever possible, use the object you are securing, such as the flanges for a railing, as a template. Have a helper position awkward objects for marking and mark each hole with a center punch so that the drill will not wander. Fasten one section at a time, so errors do not accumulate.

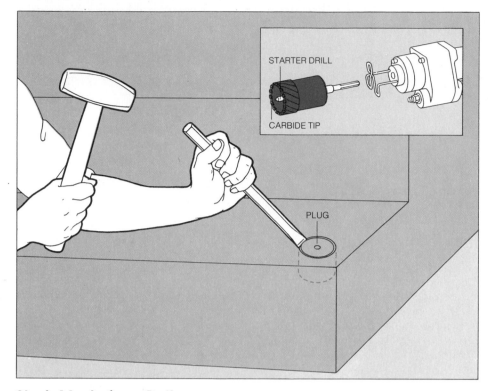

Drilling with a core bit. This carbide-tipped bit *(inset)*, which is used in a hammer drill, makes a hole by cutting a plug of masonry. The plug comes loose by itself if you drill clear through a masonry structure, but if you bore only partway you must dislodge the plug with a cold chisel. Caution: Wear goggles, hold the chisel against the edge of the plug and strike it with a 4-pound maul. If the plug does not break off cleanly at the bottom of the hole, chisel out the rest and flush away debris with a garden hose.

STARTER DRILL

CARBIDE TIP

PLUG

Studs Made from Bolts

1 **Setting a bolt in epoxy.** Mark and drill holes for the studs, then dust off the masonry near the holes. Cut a strip of 2-inch masking tape for each stud, and with a utility knife slice an X in the center of each strip to make an opening for the threaded ends of the bolts. Insert a bolt head-first into a hole and, with a putty knife, fill the space around the bolt with epoxy. Stick a piece of tape firmly to the wall, with the bolt projecting through the X; the tape keeps the bolt centered in the hole and the epoxy from oozing out. Repeat the procedure for the other bolts and allow the epoxy to cure for the length of time specified by the manufacturers before proceeding.

2 **Marking stud holes in a wood strip.** Hold a level against the edge of the board you are mounting and push it against the ends of the studs. Plumb the board with the level—or level it if it runs horizontally—then strike it sharply with the rubber face of a mallet, opposite the studs, to make slightly indented guide marks. Drill holes centered on these imprints and fasten the board to the studs with washers and nuts.

A Gun for Driving Nails

Hardened-steel masonry nails are often used to fasten wood strips to walls for paneling or shelves and to anchor wall framing to a concrete floor. A fast and easy way to do this is with a tool called a low-velocity stud driver, a special kind of gun that employs the explosive force of a .22-caliber blank cartridge to force a nail into mortar, block and poured concrete. The stud driver will not fire unless it is pressed firmly against a wall or floor. When used correctly, as illustrated in the sequence of steps shown here, a stud driver is safe, has virtually no recoil and makes very little noise. To make certain that its operation is understood, agencies that supply the stud driver usually show their customers precisely how to use the gun before they rent it.

Blank cartridges for the stud driver come with a variety of powder charges, or loads, for different jobs. A smaller charge is needed to drive a nail into mortar, for example, than into concrete, a much harder material. Choose a nail for the stud driver that is long enough to pass completely through the board you are nailing and about 1 inch into the masonry; deeper or shallower penetration weakens the fastening. For safety, keep the stud driver, nails and blank cartridges out of children's reach.

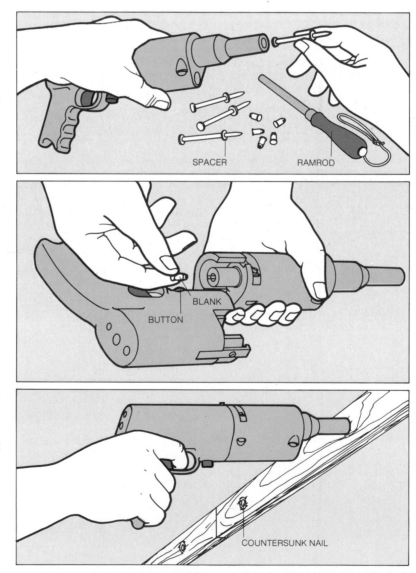

1 Loading the nail. To avoid accidents wear goggles when firing a stud driver and follow exactly these instructions for using it. First, hold the tool in one hand and with the other drop a nail head-first down the barrel. Then, using the ramrod supplied with the driver, push the nail into the barrel as far as it will go. The spacer near the point of the nail centers it in the barrel and reinforces the head once you have driven the nail.

SPACER   RAMROD

2 Loading the blank cartridge. Open the stud driver, slip the blank into the firing chamber and then close the driver. On this model, the tool is opened by depressing the button near the trigger guard and twisting the two halves of the assembly apart.

BLANK
BUTTON

3 Firing the driver. Push the barrel of the stud driver against the material you are fastening—in this case a furring strip. The barrel will slide into the driver about an inch, releasing the trigger safety. Keeping up the pressure, hold the driver at a right angle to the wall and squeeze the trigger. The driver will fire, countersinking the nail. If the nail-head protrudes from the board, re-load the empty driver with a lighter charge and fire again to seat the nail firmly, then switch to a heavier powder charge for the next nail; if the first nail is countersunk too deeply, switch to a lighter charge.

COUNTERSUNK NAIL

Cleaning Up after the Work Is Done

Masonry cleanup usually consists of at least two different tasks: disposing of mortar or concrete that remains unused at the end of the job, and then cleaning your tools.

Leftover concrete or mortar can be a nuisance, for most sanitation departments will not haul away such waste. Generally you have to take it yourself to the nearest dump. For easier handling, pour it into paper bags, or pile it in small heaps on sheets of paper and let it set into manageable lumps. Or mold concrete and save it for future use—some householders, for example, keep simple 2-by-4 wood forms ready for pouring excess concrete while it is still workable, and cast stepping stones.

Cleaning tools is an ongoing task that must be done every day. Put all your tools in a wheelbarrow, and hose them down together. Empty the dirty water into a street or road—do not let it run off into a drainage system or backyard, for even a small amount of hardened cement can clog drain traps and ruin lawns. (A town sewage system, which has large drain pipes and no traps, can carry the water off safely.) When the entire job is finished and you are ready to put tools away, wash them and coat them with petroleum jelly.

A power concrete mixer must be hosed out at the end of each workday. (Many rental companies charge an extra fee for a mixer that comes back dirty.) If you cannot clean the drum completely with a hose, turn the mixer on and pour in a mixture of water and two shovelsful of aggregate to scour it out. Empty the mixer after three or four minutes, then hose it out again. If you have waited too long to clean the mixer, you may have to scrape bits of hardened concrete out with a wire brush or chip them off with a chisel. A dull-gray exterior film of dry cement should be sponged off with vinegar.

Removing Stains and Blemishes

Ordinary scrubbing gets most blemishes off masonry, but some need special treatments with chemicals from a drugstore or garden-supply center. Caution: These substances are hazardous and many are corrosive. Follow dilution instructions, wear goggles and rubber gloves and provide ample ventilation.

Concrete and Concrete Block

□ EFFLORESCENCE (a white, powdery deposit produced when internal moisture dissolves the salts in masonry). Wash the surface with water, using a hose outdoors and buckets indoors. If water fails, apply a weak (1-to-10) solution of muriatic acid with a brush, a small area at a time, scour with a stiff brush and flush with water.

□ OIL AND GREASE. Apply a degreaser, available at auto-supply stores in powder, liquid and aerosol forms. The more costly aerosols are easier to use: spray, let stand for the time recommended, then wipe off.

□ PAINT. Soften first with paint remover, then scrape off.

□ RUST. Scrub with a stiff brush and a solution containing 1 part sodium citrate, 7 parts glycerine (both available at drugstores) and 6 parts lukewarm water. Rinse thoroughly.

□ SMOKE. Scrub with bathroom and kitchen cleanser. Rinse thoroughly.

Brick

□ BROWN STAIN (caused by the manganese used to color dark brick). Wet the brick and brush it with a solution of 1 part acetic acid, 1 part hydrogen peroxide (both available at drugstores) and 6 parts water. For a slower but simpler treatment, use a proprietary manganese cleaning compound sold under such a trade name as Brick Klenz, following instructions on the container.

□ EFFLORESCENCE. For dark brick, use the 1-to-10 muriatic-acid solution recommended for concrete. Do not use this solution on light-colored brick; it can leave stains of its own. Instead, dilute it further to a strength of 1-to-15.

□ GREEN STAIN (caused by the vanadium salts inside the brick). Wash the brick with water, then use the standard 1-to-10 muriatic-acid solution on dark brick or a weaker 1-to-15 solution on light brick. Next, wash with a solution of ½ pound of potassium or sodium hydroxide diluted in 1 gallon of water. Let the solution stand for two or three days, then wash it off with clear water.

□ MOSS. Apply ammonium sulfamate, sold at garden-supply stores, following the instructions on the package.

□ OIL AND TAR. Brush with a commercial emulsifying agent, sold under trade names such as ND-150, Clix and Big Red; for difficult tar stains, add kerosene to the agent. Hose off with clear water.

□ PAINT. Try a commercial paint remover, remove the paint with a scraper and wire brush, then wash with clear water. If the paint remover cannot be worked into all the crevices of rough-textured brick, the surface may have to be sandblasted professionally.

□ RUST. Spray or brush the brick with a strong (1 pound per gallon of water) solution of oxalic acid. Hose off the brick.

□ SMOKE. Use the method given above for concrete.

Stone

□ GRANITE AND BLUESTONE. Wash with a gentle laundry detergent.

□ LIMESTONE AND SANDSTONE. Avoid detergents; instead, use clear water and a fiber scrub brush.

□ SLATE. Wash with laundry detergent.

2

The Magnificent Mud

Cutting a control joint. Guided by a straight piece of wood, a bronze groover incises a line called a control joint into newly poured concrete. If the concrete later cracks because of heaving or settling of the earth beneath the slab, the cracks will tend to occur in the control joint, where they will be less visible—and less damaging—than if they are spread about the entire surface.

Concrete is a misunderstood material. A gooey liquid when poured, it solidifies as it dries into one of the most durable of all building materials. Prosaic as mud (from which it was originally made), it has inspired outstanding architects to create plastic fantasies—half-sculpture, half-structure. For the amateur, it is no less versatile. It can be molded into any shape from a slab to a cylinder, steps to a birdbath. And once understood, its techniques are easily mastered.

Concrete nearly always requires forms or reinforcement, and often both. Forms are generally bottomless wood boxes into which wet concrete is poured and held until a complex chemical reaction among its ingredients hardens and cures it into man-made stone. Although most forms are temporary installations, planning and building them can be the most time-consuming aspect of concrete work. For some jobs—like the steps on pages 68-69—you will need less than a half hour to pour the concrete; but plan to spend a day or two constructing and positioning the forms.

If forms guarantee that your concrete follows a desired outline, shape—and even grade—reinforcement insures that your structure holds together. Concrete's one major shortcoming is a lack of tensile strength—although it supports great weight pressing against it, it is relatively easily pulled apart. Reinforcement remedies this lack. In small slabs for sidewalks, control and expansion joints supply a kind of reinforcement. The joints channel the stresses caused by expansion, as well as by concrete's natural tendency to spread, or creep, so that cracking occurs where it causes the least damage. But for large slabs or thick blocks, it is necessary to add steel reinforcement in the form of heavy wire mesh or steel rods. Because concrete dries quickly, careful planning is essential. During the crucial 45 minutes after pouring, work rapidly so that the concrete does not begin to harden before you have completed the several smoothing operations required. If you want to pour a large project, such as a driveway or patio, all in one swoop, you will need helpers. But you can handle it alone, at your own pace, if you divide it up into small sections, as described on page 61.

Whether you assemble a task force or work by yourself, there is no need to settle for utilitarian drabness. Practical and durable the structure will be if you simply follow the recipes. Attractive it can also be if you apply your own imagination. A sidewalk, for instance, need not be a dull gray series of rectangles; it can be shaped to any pattern, given color with pigment, or texture with stones pressed into its surface. You can do the same with steps, a driveway or a patio, for the plasticity of concrete makes it the most versatile of masonry materials.

Building with Concrete: The Preliminaries

Almost any concrete work you undertake —even a patio tucked into a corner of your yard—may run up against state or municipal regulations or even restrictions in the fine print of your house deed or mortgage. Before you start, check with local authorities. Many building codes and zoning laws dictate the dimensions, location and design you can have, and the standards your materials must meet. You may need a building permit.

Rarely do these technicalities place any unreasonable burden on you; in fact, they often can help you by preventing mistakes. What you must contend with is nature: the character of the site where you want your project to go. Take into account the slope of the terrain, the rock outcroppings and the ponds or streams, the trees, shrubs and their roots, the location of gas, electric, water or sewage lines and of dry wells, septic tanks or cesspools—including abandoned ones. Measure the distances that will be important in planning your project. Then draw a scale layout that shows existing structures and landscape features.

At the site, dig a test hole about a foot deep to check the soil. Concrete paving and a footing for a garden wall do not need to extend below the frost line as house foundations do, but they require excavations deep enough to provide space for a gravel drainage bed under the concrete. The bed keeps the concrete dry and permits it to shift, without cracking, when the earth heaves with changing weather. If your soil is sandy, a 2- to 4-inch bed will be adequate for a 4-inch concrete slab. If your soil is clayey or rocky you need a 6-inch bed.

Whatever the soil, the ground must be stable. Avoid any site with more than 3 feet of recent landfill, which might settle. And if you strike water in the test hole, or if you live in an earthquake zone, the job requires professional skills.

After you make your test to decide how deep the excavation should be, you can clear the area by moving plants that might be in the way and, if necessary, getting rid of old concrete. To break up concrete that is not reinforced with wire mesh, lift and drop a heavy sledge—do not try to hammer—working from the edges toward the center. For reinforced concrete, call in a professional. If you live in a cold climate, you can break unreinforced concrete during the winter by drilling 1½-inch-deep holes into it at 1-foot intervals with a ½-inch masonry bit. Fill the holes with water during a cold spell; the water will freeze and burst the concrete. An old sidewalk will break if you lift up an edge with a crowbar, slide a rock underneath and then pull out the bar so the slab falls on the rock. Dispose of the concrete chunks by burying them deeply or hiring a hauler to take them.

With the site cleared, check your preliminary drawings by using garden hose or string to lay out the sidewalk or other project in the place where you plan to build it. Then, on the basis of the new measurements, you can decide whether to make your own concrete (pages 26-29) or buy it ready-mixed (right). Since you will divide most slabs into segments to permit them to expand and contract as the temperature fluctuates, it is a simple matter to install home-mixed concrete one segment at a time, a procedure that permits you to prepare small batches yourself. For faster results you may order large quantities delivered by a transit-mix truck.

Estimating how much concrete you need for the project, and how much gravel or sand the drainage bed requires, takes only simple arithmetic. All three materials are sold by the cubic yard so you must estimate the length, width and depth dimensions in advance. For slabs, use the table below to figure out quantities. For footings and steps, multiply the length by the width by the depth to determine the total cubic feet. There are 27 cubic feet in a cubic yard, of course, but you may find it simpler to divide the cubic feet by 25—a trick that automatically gives you cubic yardage with an extra allowance of about 8 per cent for waste.

Estimating Materials for Concrete Slabs

| Area of slab | Thickness of slab | | | | | |
	1 in.	2 in.	3 in.	4 in.	5 in.	6 in.
10 sq. ft.	.03 cu. yd.	.06 cu. yd.	.09 cu. yd.	.12 cu. yd.	.15 cu. yd.	.18 cu.yd.
25 sq. ft.	.08 cu. yd.	.16 cu. yd.	.24 cu. yd.	.32 cu. yd.	.4 cu. yd.	.48 cu. yd.
50 sq. ft.	.16 cu. yd.	.32 cu. yd.	.48 cu. yd.	.64 cu. yd.	.8 cu. yd.	.96 cu. yd.
100 sq. ft.	.32 cu. yd.	.64 cu. yd.	.96 cu. yd.	1.28 cu. yd.	1.6 cu. yd.	1.92 cu. yd.
200 sq. ft.	.64 cu. yd.	1.28 cu. yd.	1.92 cu. yd.	2.56 cu. yd.	3.2 cu. yd.	3.84 cu. yd.
300 sq. ft.	.96 cu. yd.	1.92 cu. yd.	2.88 cu. yd.	3.84 cu. yd.	4.8 cu. yd.	5.76 cu. yd.
400 sq. ft.	1.28 cu. yd.	2.56 cu. yd.	3.84 cu. yd.	5.12 cu. yd.	6.4 cu. yd.	7.68 cu. yd.
500 sq. ft.	1.6 cu. yd.	3.2 cu. yd.	4.8 cu. yd.	6.4 cu. yd.	8.0 cu. yd.	9.6 cu. yd.

Matching area, thickness and cubic yards. This table gives the number of cubic yards of concrete, sand or gravel required to make concrete slabs and the drainage beds beneath slabs up to 6 inches thick—the thickest an amateur is likely to attempt. To use the table, you must first settle on the thickness of the slab or bed and then calculate its area in square feet. Look up the yardage required by locating the horizontal column that lists an area closest to yours and going across to the vertical column for the thickness you need. A slab 4 inches thick and 400 square feet in area, for example, calls for 5.12 cubic yards of concrete. Add about 8 per cent allowance for waste and spillage.

Coping with the Ready-Mix Truck

For large jobs you can save work by buying the concrete from a transit-mix company that will make it to your specifications and deliver it ready to be poured into place. But you may have to work fast and enlist several assistants. Tell the dealer how many cubic yards you need *(bottom left)* and describe how you want to use it. In his jargon, most jobs will call for 5½- to 6-bag concrete—made with about 520 to 560 pounds of cement to the cubic yard. Where winters are severe, use 6-bag concrete, elsewhere 5½-bag.

Professionals measure the consistency of wet concrete by the number of inches a cone-shaped mass will slump when the conical form is lifted off. You will need concrete with a 4- to 6-inch slump. Concrete with less than 4 inches of slump is too stiff to be workable; with more than 6 inches it becomes soupy. Also, ask for a coarse aggregate —gravel—with a maximum size of 1 inch

diameter and for a 6 per cent air entrainment *(page 26)* in freezing climates, 4 per cent elsewhere.

How much you pay for the convenience of ready-mix depends on the size of your order. If you need more than 3 to 5 cubic yards of concrete, you may find the mix costs less than buying materials and combining them yourself. If you need less, you may have to pay a delivery charge or a higher price per cubic yard. Most prices include 30 to 45 minutes delivery time. When the truck spends longer than that unloading, you pay an hourly rate for the extra time.

To speed up delivery, decide in advance where the truck will stop and how you will get the concrete to the work site. Usually the best plan is to have the truck park on the street and use chutes and wheelbarrows to unload the concrete. A 10- to 12-foot chute is standard equipment, and most dealers can supply a 12-foot extension chute.

Trying to deliver concrete more than 24 feet by chute is not practical; at greater distances, use wheelbarrows and lay a piece of plywood under the bottom of the chute to catch the drippings. Then line up planks to form a path from the truck to the forms—a wheelbarrow full of concrete is too hard to push over a soft surface like lawn. Organize your crew of helpers so that some are placing fresh concrete while others smooth and finish.

Having the transit-mix truck drive up onto your property is rarely advisable. The weight of the truck can break a curb, driveway or sidewalk and make ruts 3 feet deep in a lawn. If the time and work saved would be worth the risk, however, you can minimize the damage by laying 2-inch-thick planks along the truck's route to equalize the load. Never let the truck drive over septic tanks or cesspools; it could crash through the ground.

Unloading ready-mixed concrete. Place a 4-by-8 plywood panel where the truck will discharge the concrete. Then lay 2-foot-wide plywood strips or 1-by-8 planks from their unloading area to your work site to make a firm surface for roll-ing a loaded wheelbarrow—it may sink into a lawn. You may need a helper to shift these wooden tracks as you work so that you can always reach a convenient point to dump your barrow load. When the truck arrives, work quickly with a shovel to slide the concrete from the delivery chute into a wheelbarrow. Take care not to overfill the wheelbarrow. And do not leave concrete in the chute—it may drop out before you can get back to it again.

Setting Up Forms for Slabs

Forms are molds that shape concrete into slabs for sidewalks, patios and driveways, and footings for low walls and cinder-block steps. They are built like open-ended boxes. Boards and metal strips, set on edge, outline the outer edges of the concrete and may also divide a large area into smaller—and more workable—units. Except for forms made of redwood or cedar, which will remain in the slab as a permanent decorative feature *(page 61)*, forms are temporary constructions—assembled on the site before the concrete is poured and then removed after the concrete has cured.

Even for small jobs, forms have to be strong enough to withstand the weight of the concrete. Straight form sections for a simple 4-inch slab, such as a sidewalk *(bottom)* or patio, require 2-by-4s. Where you need to shape curved sections of concrete, forms can be made with 4-inch-wide, $1/16$-inch-thick sheet-metal garden edging.

Temporary forms are best made of fir, pine or spruce, which should be smooth and, unlike lumber used in most construction, green. Fully dried wood may absorb moisture from the concrete and interfere with curing. To determine how much you need, measure the outer dimensions of the marked area *(opposite, top)*. Continuous boards will make the strongest forms, but long sections may have to be pieced by nailing strips to the backs of abutting boards. Separate form boards may be necessary to support the asphalt-impregnated filler used for expansion joints within the slab *(page 49)*.

In addition to these materials, you also need a generous supply of 18-inch stakes cut from 1-by-2s or 2-by-2s. They are used first to mark the layout and then to support the forms and to brace them.

The support stakes also indicate the grade, or pitch, of the slab that directs rain water away from the nearest building. If you are building a simple sidewalk that will run in a straight line over land that slopes gently away from the nearest building, follow the slope of the land. The sidewalk then will be flush with the ground, and the stakes for forms are driven so that their tops are flush with the ground as well. If the slope is abrupt, or if the slab is large or complicated by an-gles and curves, the concrete must be pitched for drainage as shown in detail on the following pages.

Forms are installed after you have laid out, excavated and leveled the area. The depth of the hole you need to dig depends on local codes, but for simple structures like walks, patios, steps and driveways, 10 inches is generally enough. There is no need to dig down below the frost line. The slab rests on a bed of gravel and, unless it is smaller than about 40 square feet, is broken up by expansion joints into smaller units. The gravel and joints allow the concrete segments to shift independently as frost heaves the ground—the units float like rafts on a gentle sea—so that the slab will not be subjected to cracking stresses.

After the forms are assembled and graded, the gravel drainage bed can be laid in place *(page 49)*. Large slabs—patios and driveways—need a layer of wire-mesh reinforcement *(pages 58-63)*, but small slabs for walks do not. After the bed has been leveled and the expansion joint fillers set in place, the site will be ready for concrete.

The Simplest Slab

Building basic forms. A straight walk on gently graded ground requires only a little time and effort to lay out. After the walk is outlined and the site excavated and leveled, support stakes are set at both ends—1½ inches outside the proposed slab width to allow space for form boards—and at 3- to 4-foot intervals between. Pound the tops of the stakes flush with the ground level *(below)*. Nail the boards to the stakes, flush with the tops and suspended above the bottom of the excavation to provide for a 2- to 6-inch-deep gravel drainage bed beneath the slab. Bracing stakes, set diagonally, supply additional strength where form boards are pieced. If the site does not slope gently away from the nearest building, or if curves and turns are involved, the slab must be pitched as shown on the following pages.

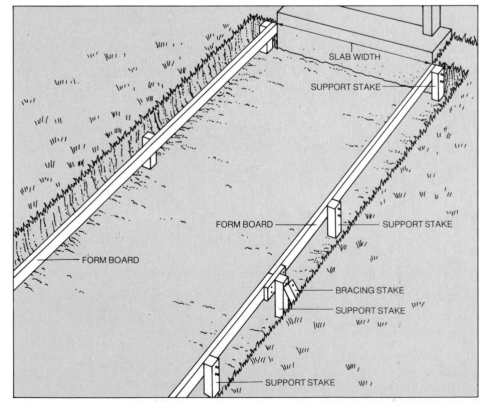

SLAB WIDTH

SUPPORT STAKE

FORM BOARD

SUPPORT STAKE

FORM BOARD

BRACING STAKE

SUPPORT STAKE

SUPPORT STAKE

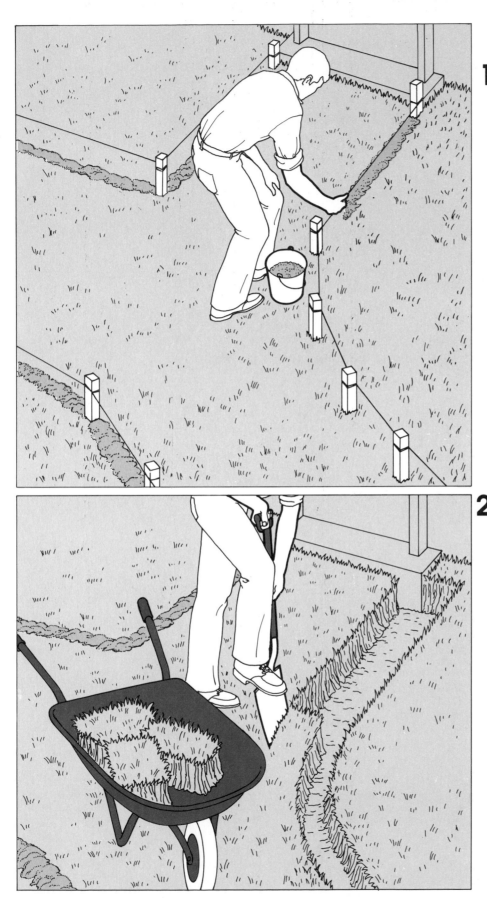

Excavating the Base

1 **Outlining the slab.** For a straight small slab, drive pairs of stakes on both sides of the proposed slab at the beginning and end of the site. Tie strings to the stakes. To mark the string line on the ground, take a handful of sand and let it trickle out over the strings to make a sand outline. For a curved slab, drive additional stakes at 1- to 2-foot intervals around each curve and run the strings for both sides around these stakes. For a forked sidewalk, drive additional stakes around curves and at the beginning and end of each straight section before attaching strings.

2 **Preparing the base.** After removing marker stakes and strings, dig a trench of the required depth—10 inches for the walk shown here—1 foot outside one sand line. Save the turf if you want to replace it later along the edge of the finished slab. Then, digging from inside the slab area, complete the excavation, extending the trench to 1 foot beyond the opposite sand line.

3 **Leveling the base.** Smooth the base with a rake and fill in with sand or gravel any holes left by the removal of large rocks. Flatten and level the earth by pulling a long 2-by-4 board across the surface. For a sidewalk or other narrow slab, use a board about as long as the base is wide.

4 **Tamping the base.** Compact the base with a lawn roller, a spare tire *(page 71)*, or a tamper made by nailing and bracing a 2-by-4, 3 or 4 feet long, to a board 1 or 2 feet square. Nail screen-door handles at the top, one on each side. Fill and tamp low spots. Then use the leveling board *(Step 3)* as a straight edge to be sure the trench bottom is fairly even across its entire width.

Staking the Site

Driving support stakes. For a straight small slab, restake the ends within the excavated base, setting the stakes 1½ inches outside the desired slab width. Drive the stakes about 6 inches deep. Run strings along the inner faces of each set of stakes. Then drive intermediate support stakes at 3- to 4-foot intervals along both sides of the slab and at the joints of boards, positioning the inner faces of the stakes against the strings.

For a curve, or fork, drive extra pairs of stakes at the ends of each additional straight section before attaching the strings. Do not set stakes around curves at this point.

On a naturally sloped site, grade the slab by pounding down each support stake until the top is at ground level. Then skip the grading instructions below and overleaf, and proceed to setting up the form boards (*page 47*).

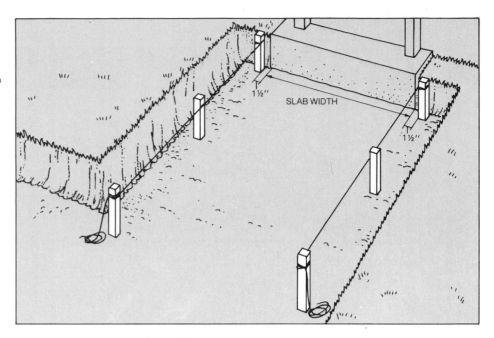

Grading the Stakes

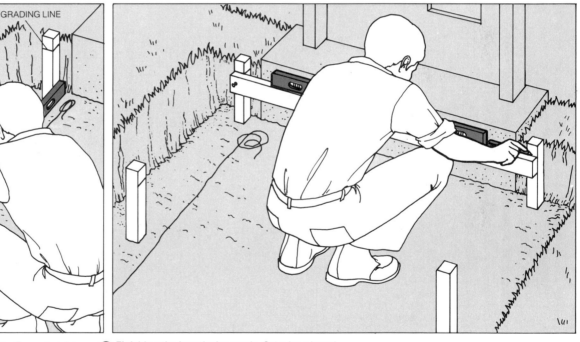

1 Setting the lengthwise grade. For a straight slab on perfectly flat ground, use a waterproof marker to draw a line at ground level on the inner face of one of the support stakes nearest the building. Continue the line around the side of the stake away from the building. Then use a level to make a mark at the same height on the adjacent stake. To slope the slab away from the building, draw a grading line ½ inch below the mark. Use the level to mark a grading line at this height on each of the adjacent support stakes along one side of the slab. For a forked sidewalk, like the one shown, treat one fork as the main section for the grading.

2 Finishing the lengthwise grade. Set a long board across the width of the slab in front of the stakes nearest the building. Position the top edge of the board under the line on the marked stake and drive one double-headed nail to hold the board temporarily in place. Then level the board and mark a line on the opposite stake. Remove the board. Indicate the lengthwise slope for the forms by drawing a ½-inch lower line on the adjacent stake, following the method in Step 1.

3 **Setting widthwise grade on flat ground.** At the far end of a straight slab or the main section of a forked walk, align a board with the grading line on the one stake that is marked. Drive one nail to hold the board temporarily. Use a level and the board to mark the opposite stake. Remove the board and draw a grading line on this second stake ½ inch below the mark. Fasten string from this grading line to the grading line on the second stake from the building. Mark the intermediate support stakes at the level where the string crosses the inner faces. Remove the strings.

4 **Finishing grading for a fork.** Grade the remaining fork widthwise only. Use the level to mark the stakes along the side nearest the building at the same height as the mark on the stake at the inner juncture of the fork. Grade the opposite side by running a string from the end stake to the stake at the outer juncture of the fork, sloping it so that it is ½ inch lower at the end of the section than at the beginning.

Nailing on the Form Boards

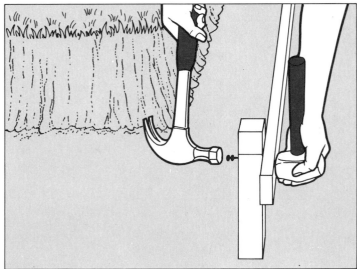

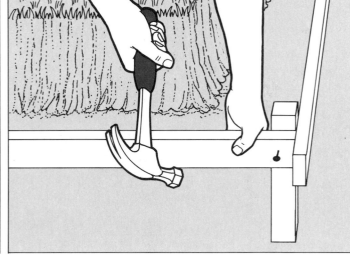

1 **Attaching form boards.** For a slab on sloping ground, drive a double-headed nail into a stake near the building, just below the top of the outer face. Lift one end of a form board level with the top, and drive the nail into it, using a mallet or sledge to cushion the blows. Attach the board to the stake at the other end, then at the center. Add a second set of nails below the first set. Finally attach the board to the intermediate stakes. For a slab on a flat site, position the top edge of the board flush with the stake's grading lines.

2 **Making corners.** Butt the end of the corner form board against the back of the board for the straight section. Then use a single-headed nail to fasten the corner board to the face of the support stake, checking to be sure that the top edges of both boards are level.

3 **Adding braces.** Drive a bracing stake angled from the form top into the excavation outside the support stakes at each corner. At joints where form boards are pieced, drive a perpendicular support stake before putting in a diagonal brace. Nail braces to support stakes below existing nails.

4 **Trimming forms.** For a forked sidewalk, saw off the ends of form boards where they overlap corners. Trim the tops of support stakes that project above form boards—except the pairs on the ends of curves—to provide a uniform flat surface for screeding the slab later.

Shaping Curves with Metal

1 **Staking curves.** Arrange string in the desired arc around a curved section of the slab. To align the face of the 1/16-inch-thick sheet-metal strip for the curve with the face of the 1½-inch-thick form boards, begin and end the arc 1½ inches inside the adjacent form-board stakes. Drive support stakes for the curve at 1- to 2-foot intervals along the string line. To grade the curve for a slab on flat ground, run the string from the top of one form-board stake to the other across the faces of the stakes. Adjust the string so it is level with the tops of the adjacent form boards and pull it taut. Use a marker to draw grading lines where the string crosses each stake. Remove the string.

2 **Attaching curved forms.** Cut a 4-inch-wide strip of sheet-metal garden edging about 6 inches longer than the curve. Attach the strip to the form board at one end of the curve, with ½ inch galvanized nails, overlapping the board by about 3 inches and aligning the top of the strip with the top edge of the board. Continue nailing around the curve to the form board at the opposite end. Add diagonal bracing stakes at each support stake, then saw off the tops of projecting support stakes level with the boards and metal strip. Backfill the trench outside the curve with earth for extra support, but be careful not to tamp the earth so hard that you move the stakes.

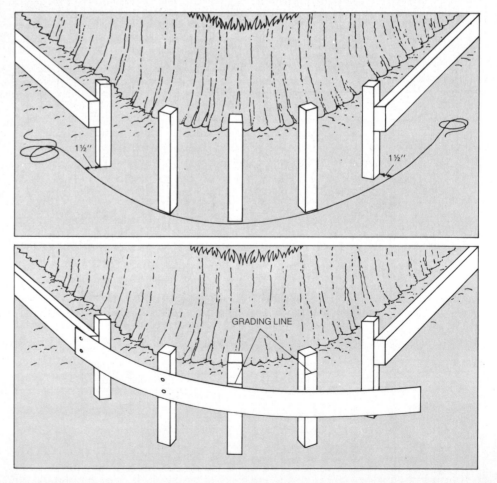

Installing a Drainage Bed

Placing gravel. Place a 6-inch-deep layer of gravel inside the forms. To level the gravel even with the bottom of the form boards, pull a screed —a 1-by-4 or 2-by-4 placed on edge—toward you across the gravel surface. Make the screed the width of the space between the form boards and attach a wider 1-by-4 or 2-by-4 to the top so you can rest the projecting ends on the forms as you pull the screed. Fill in low places with gravel as you go. Use the tamper (*page 44, Step 4*) to compact the bed, adding more gravel if necessary. Then level the surface again with the screed.

Expansion Joints

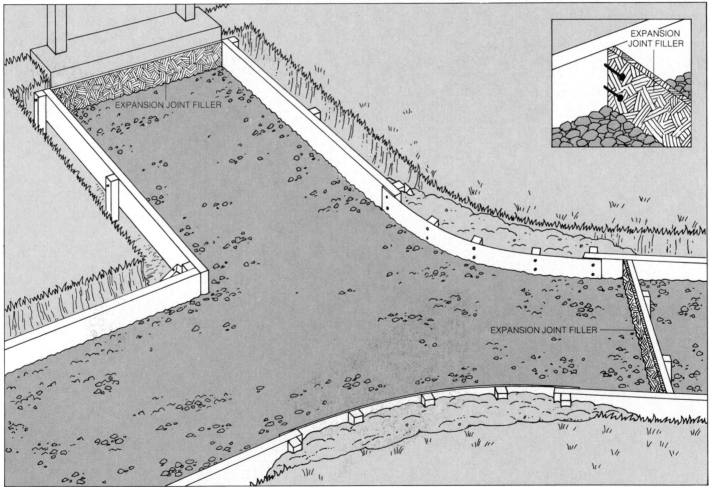

EXPANSION JOINT FILLER

EXPANSION JOINT FILLER

EXPANSION JOINT FILLER

Installing joint filler. To allow for heat-caused expansion and contraction in a slab, flexible joints are needed wherever the slab abuts an existing rigid structure, such as a step, wall or other slab. Cut pieces of 4-inch-wide asphalt-impregnated expansion joint material to the width of the slab. Place a piece between the slab and any abutting structure. (The semirigid material will stay in position by itself.) Within the slab, expansion joints are needed every 10 feet in a 3- to 4-foot-wide sidewalk, every 8 feet in driveways and every 8 to 10 feet—both lengthwise and widthwise—in patios. If you plan to pour the concrete one section at a time, you can support these strips of filler material by placing a board across the slab at a right angle to the forms and staking the board from the outside. If you plan to pour all of the concrete for the slab at once, stand the filler material in place and support it on both ends by driving pairs of galvanized nails partway into the form boards at the front and back of the filler strip (*inset*). Then, when you pour the concrete into the forms, mound a little on each side of the strip to keep it upright while you fill the forms.

Pouring and Finishing

Speed is the most important requirement for placing concrete into forms. A fresh mix needs to be poured and finished in about three hours; after that it is too hard to be workable. In that time, one man, working alone with transit-mix concrete, can place and finish about 30 feet of a 3-foot-wide sidewalk like the one shown here. One assistant will let him finish 50 feet of walk in the same time; two assistants, 70 feet.

As soon as concrete is poured into place *(bottom)*, it must be compacted and leveled to the tops of the forms—first with a spade, next with a screed, a long board that is pulled along the surface, and then with a trowel-like darby. Compacting the concrete forces water to the surface. When this bleed water appears, all work must stop. Darbying or troweling before bleed water has completely evaporated can make the slab crumble later. Air-entrained concrete bleeds relatively little, but even with it, 10 to 20 minutes may be required for the water to evaporate completely on a hot dry day . —and an hour or more may be needed in cool or humid weather.

Once the sheen of the bleed water disappears, the surface must be finished and the edges rounded to reduce the chances of chipping. For a slab you must also incise control joints into the surface every 4 or 5 feet to induce cracks at the joints rather than at random. If you want the marks of the edger and jointing tool to be visible, finish the surface with the desired texture first. Otherwise, round edges and joints before finishing.

Concrete can be colored by mixing pigment into the entire batch. It can also be colored during finishing by dusting the top with a dry-shake mixture or by applying a 1-inch-thick layer of colored mortar on top of a normal concrete base. And, of course, it can be colored with paint or stain, providing you let it cure for at least six weeks and use specially formulated concrete coatings.

For dusting with color, you can buy dry-shake or make it yourself with two parts of portland cement, two parts of mortar sand and one part of cement pigment. A pound covers about two square feet. Apply the dry-shake—both before you finish the surface and again after-ward—by leaning over the wet concrete and letting the mix sift through your fingers. For a mortar coat of color, pour unpigmented concrete up to 1 inch below the top of the forms and screed it level. As soon as the concrete begins to dry, apply the colored mortar and screed it level with the top of the forms. Any desired finish except the smoothest may be applied to precolored concrete; a grooved or flagstone texture can be given to color-dusted or mortar-coated concrete. Use wooden finishing tools; a metal tool may discolor the surface.

Concrete must be kept damp and undisturbed to cure for at least a week so that the chemical reactions that give it strength can proceed uninterrupted. After curing, the forms can be removed, but heavy loads must be kept off the concrete for an additional week.

Placing the Concrete

1 Filling the forms. Hose the forms and the gravel bed thoroughly, but leave no puddles. Dump the first wheelbarrow load of concrete into the forms—those farthest from the truck if you are using transit-mix. Then pack each successive load up against the preceding one, overfilling the forms by about ½ inch. Be careful not to knock down or cover up the expansion joint filler. If you do not support the filler strips within the slab with form boards, as shown here for section-by-section pouring, mound wet concrete on either side of each filler strip.
If you plan to add a topping of colored mortar, fill the forms only to within an inch of the tops. As you go, shovel the concrete into the corners of the forms, and jam the shovel edge through the concrete to eliminate air pockets.

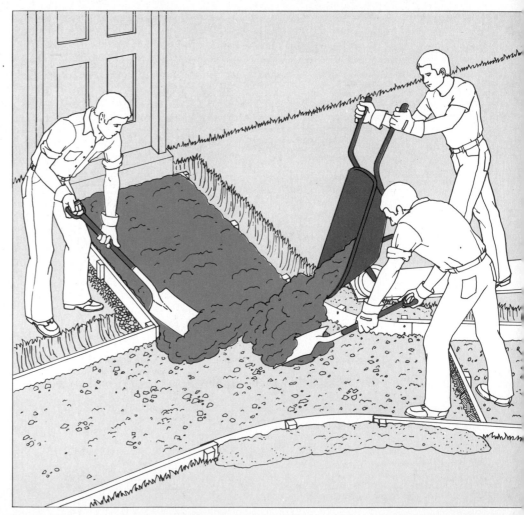

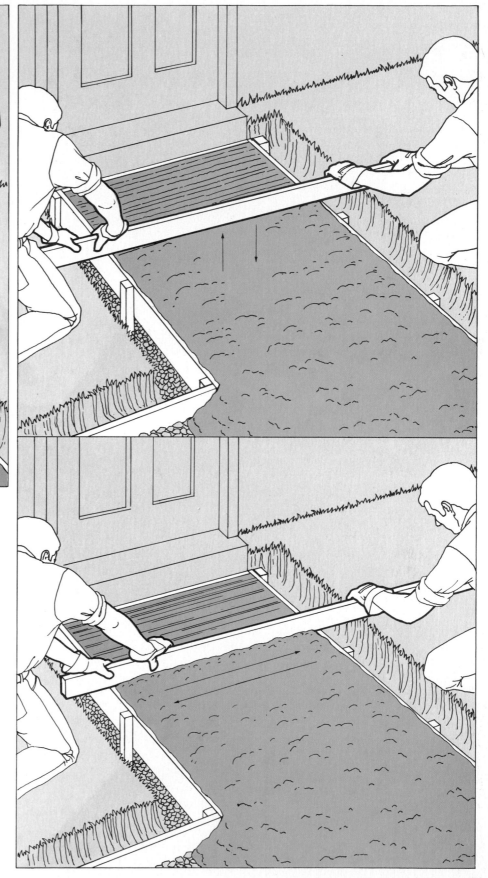

2 **Spading the edges.** As soon as one section is completely filled with concrete, drive a flat spade down between the concrete and the inner edges of the forms to force large aggregate away from the outside of the slab.

3 **Striking off the concrete.** Compact and level the slab by using the edge of an unwarped 2-by-4 or 2-by-6, about 2 feet wider than the forms, to strike off or screed the freshly poured concrete. Lift and push down (above, right) to force large aggregate down. Then zigzag the board (right).

To cover concrete with colored mortar, make a screed of two pieces, one to fit down between the forms 1 inch, the other to slide along the tops of form boards. When the surface looks dry, pour in mortar and screed again—this time level with the tops of the forms. Then proceed as for uncolored concrete.

If you plan to embed stones in the surface of the slab, complete the level-screeding process and then skip to the instructions on page 55.

4 **Floating the surface.** Working quickly, smooth the concrete and eliminate any "hills" or "bird-baths" with a darby—a board about 4 inches wide and 3 feet long with a handle attached. Pressing down lightly on the trailing edge of the darby, sweep it back and forth across the surface in wide arcs to force large aggregate down into the concrete. When bleed water floats to the surface, stop darbying and wait until the water evaporates before either finishing the surface or edging and jointing the slab.

If you plan to create a flagstone finish, complete the floating process and then skip to the instructions on pages 54 and 55.

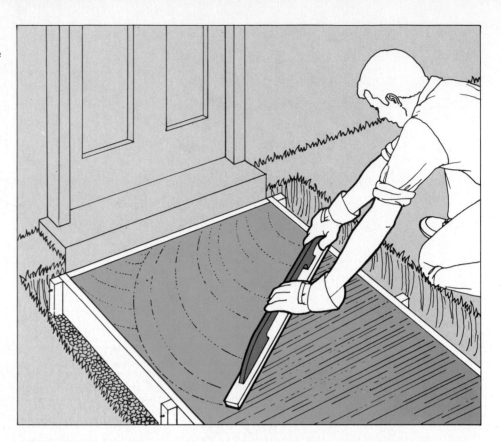

The Final Touches

1 **Finishing the surface.** As soon as bleed water evaporates and the shiny surface turns dull, smooth and compact the concrete with a wood trowel. Keeping the trowel pressed flat, sweep it back and forth in gentle curves. Support yourself on a second trowel if you need to lean over the concrete. If the slab is too wide to reach across, place two boards over the forms so that you can kneel on them. Move backward as you work, troweling out any marks you might have made by leaning or kneeling on the concrete.

If you plan to color the concrete with a dry-shake, dribble the mixture evenly over the surface as soon as it turns dull and again after troweling.

Caution: Do not overwork the concrete. Too much troweling will bring water and cement to the surface, weakening the slab.

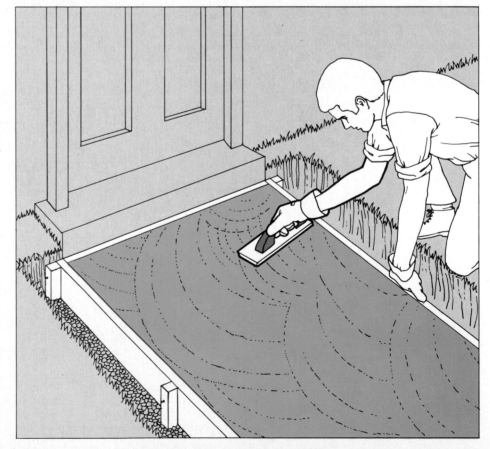

2 Rounding the side edges. Draw a small trowel along the inner edges of the forms *(near right)* to cut the top inch or so of concrete away from the wood. Then finish the sides of the slab by running an edger *(far right)* firmly back and forth until the edges are round and smooth.

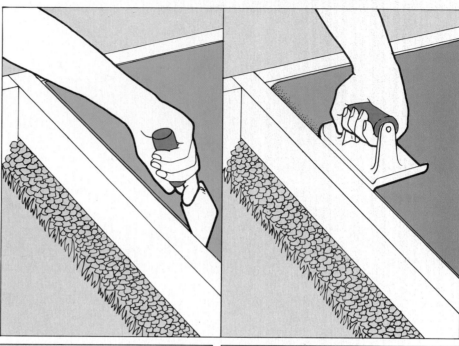

3 Cutting control joints. Make a ¾-inch-deep control joint across the slab every 4 feet. First place a guide board, such as a 2-by-8, across the forms; check with a square to make sure it is perpendicular to the outside edges of the slab. Then force a jointing tool down into the concrete along the edge of the guide board. Run the tool back and forth to round the inside of the joint.

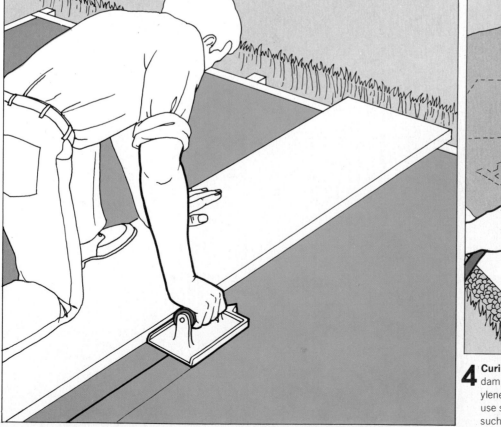

4 Curing the slab. Wet the concrete and keep it damp for a week. Cover it tightly with polyethylene sheeting anchored with bricks; you can also use soaking-wet burlap sacks or old rugs, but such coverings must be kept damp with a hose.

Special Finishes and Textures

A textured surface or design can add visual interest—and a safer foothold—to a concrete slab. The surfaces in these photographs are easily applied using the techniques sketched.

Two of the finishes incorporate safety features into their design: embedding stones or making grooves in the surface will improve its tread. The simulated flagstone effect and the highly smooth surface, on the other hand, serve to provide decorative veneers for your sidewalk or patio area.

Troweling a smooth surface. To finish a slab with a silky surface, wait until bleed water has evaporated and work the surface with a wood trowel. Then work it again one, two or even three times with a steel trowel, depending on the degree of smoothness desired. For the first troweling keep the blade nearly flat against the concrete, and sweep the trowel back and forth in arcs 2 to 3 feet wide. Wait a few moments to let the concrete harden slightly between trowelings. Use more pressure and a successively higher lift to the front edge of the blade (drawing) for each additional troweling. With each troweling, the recognizable trowel marks on the surface will become fainter and will have almost disappeared after a third troweling. Work until no concrete collects on the trowel and the tool makes a ringing sound as you sweep it across the slab.

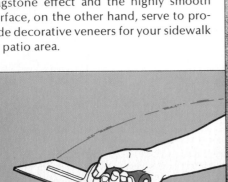

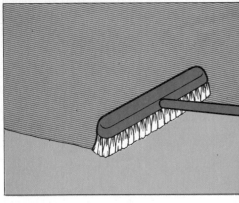

Brushing on a nonskid surface. To give concrete a safety-tread surface like that in the photograph at right, draw a stiff-bristled broom across the concrete after you finish working it with a wood trowel. Sweep straight lines at right angles to the sides or, if you want a curved pattern, move the broom in arcs. If the broom kicks up small lumps of concrete, hose the broom clean and let the slab dry a few minutes longer before you proceed. If you have to press down heavily to score the surface with the broom, work fast; the concrete will soon be too hard to take this finish.

Embedding stones. Multicolored stones inlaid in concrete create an attractive, skidproof surface. Round stones, ¾ to 1¼ inch in diameter, called aggregate, are available at masonry suppliers, who will deliver the proper amount for your slab. Wet the stones so they will not absorb water from the concrete and spread them over the slab after screeding. Stones can be dropped randomly to create a terrazzo effect (below) or placed in circles, squares or any other design you wish. Tap the stones down with a darby until the tops are just under the concrete surface (below, upper right). If they sink out of sight, wait for the concrete to firm. If you cannot force the stones down with the darby, use a brick. Then level with the darby. After bleed water disappears, place a board over the slab. When you can stand on the board without pushing the stones farther down, brush the concrete until the top of each stone projects from the surface. Flush away excess concrete with fine spray from a hose (below, lower right). Do not expose more than the top third of the stones. Cover the concrete to begin curing; after a day or two, uncover it and wash the stones with 1 part commercial-strength muriatic acid in 10 parts water. Caution: Wear goggles and rubber gloves. Cure for another 5 or 6 days.

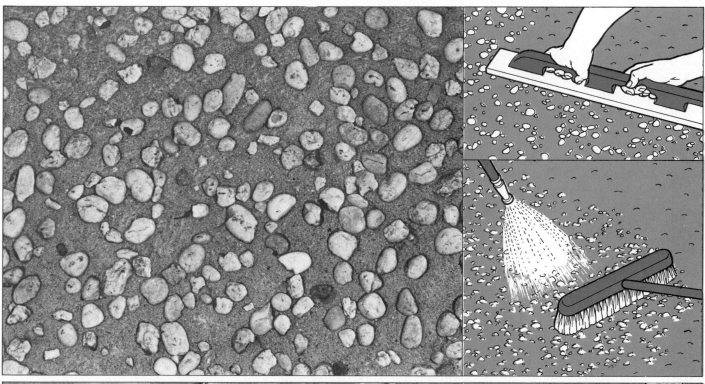

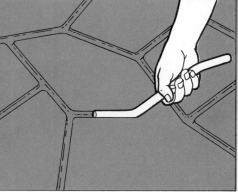

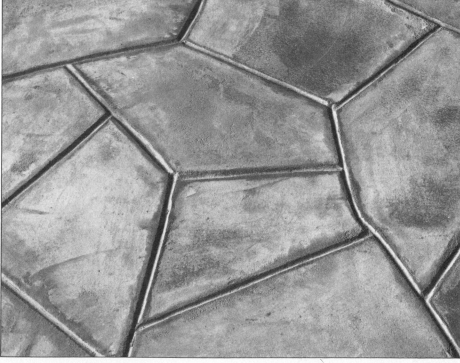

Imitating flagstones. After you have completed smoothing the concrete with a darby, carve irregularly spaced grooves ½ to ¾ inch deep into the wet surface. A convex jointer is an excellent tool for this purpose, but you can also use a length of copper pipe bent into an S-shape.

When the bleed water evaporates, trowel the concrete and retool the grooves until the flagstone-like pattern shows distinctly. Then brush out the grooves carefully with a dry paint brush to clean out loose bits of concrete that remain.

55

What Goes Wrong and Why

When a concrete slab is correctly poured and finished it is nearly indestructible, but faults in preparation, working or finishing can lead to the defective surfaces shown in these photographs.

Over-troweling, or darbying, is a primary cause of concrete failure. Such overworking sends aggregate toward the bottom of the slab and brings too much water and cement up. If the top of the slab contains too little aggregate, which provides strength, the slab surface will break up.

Inadequate curing is a second problem. If a slab dries out or freezes and thaws too soon after pouring, severe surface damage will result.

The third common cause of defects is an incorrect mixture. Too much cement or water will weaken the entire slab. If you use aggregate containing soft stones or clay lumps instead of hard gravel, the surface may break down under normal wear and weather.

Dusting. The gray powder on the finger at left is picked up by a touch on a surface that has been overworked with a darby or trowel. Too much working weakens the top layer by dividing the concrete mixture—heavy aggregates sink to the bottom, leaving too much water and cement on top. These light elements are not strong enough to withstand normal wear. Dusting may also occur if the slab has been cured less than a week.

To stop dusting, brush the surface thoroughly to remove loose material. Then apply concrete sealer, which is an acrylic-polymer compound available in gallon cans at most paint stores. A gallon will cover about 400 square feet.

Spalling. Reinforcing rods, placed too near the surface, are bared by the flaking away of the thin layer of concrete over them. Such spalling also occurs in unreinforced slabs if the surface is weakened by too much darbying or troweling.

You can repair a spalled concrete surface by patching it with a sand-cement-epoxy mixture, as illustrated on pages 20-22.

Scaling. A concrete mixture containing too much water will lack strength throughout when it dries, and the weak top layer may crumble. Scaling also occurs if the slab is subjected to freezing and thawing before it has a chance to cure properly.

Popouts. If the aggregate contains lumps of clay or crumbly stones, these soft elements will deteriorate and wash away when the concrete has dried, leaving ugly holes in the surface called popouts. Make sure that the aggregate you use contains only hard gravel to avoid this defect.

Cracking. Large cracks like these open up in concrete that contains too much water and cement, or that was poured so rapidly it could not be compacted properly. Spade freshly poured concrete thoroughly to force it into all corners of the form. Do not let a batch of concrete dry before pouring the next batch against it: cracks may appear at the boundary between the two pours.

Crazing. A network of hairline cracks may spoil the surface of a concrete mixture that contains too much cement; such an over-rich mixture shrinks more than a normal one when it dries, which leads to cracks. Lack of sufficient curing will also cause crazing.

For a Large Slab: Reinforcement

All the techniques for building a small concrete slab, such as a sidewalk, are used in making a large one for a patio, a driveway, a play area. There is one addition: a reinforcement of wire mesh. It is needed to help hold the concrete together against cracking under the heavy loads—often in opposite directions—that a big slab may get, and also to resist "creep," the natural spreading that is negligible in small sections but adds up in big ones.

Specific requirements for reinforcing concrete may be included in your local building code. Therefore, before you start one of these large projects, be sure

to check with your building department.

Wire reinforcing mesh, which looks like heavy fencing, comes in rolls 5 feet wide and 150 feet long, and is sold by the square foot. For a patio, buy a size called 6×6-10/10, meaning the wires are 10 gauge, spaced 6 inches apart in each direction. Heavier gauge, 6×6-8/8 or 6×6-6/6 is needed for driveways, steps and barbecue foundation slabs.

Building material suppliers sometimes charge a high fee for cutting the mesh to size, and you may save by buying a whole roll and cutting it yourself, even if some is left over.

Cut reinforcement 2 inches smaller

than the forms, building up sections more than 5 feet wide by overlapping pieces 6 inches and binding them together with thinner wire. As you complete preparations for pouring, walk over the mesh to flatten out the waves formed while it was in roll form. Wear gloves to protect your hands.

With several workers to help with the finishing steps, you can pour a fairly large patio in one day, using ready-mix concrete from a truck. But if you prefer to mix your own concrete and pour it in small batches, construct a patio with permanent forms (page 61) that are filled one at a time.

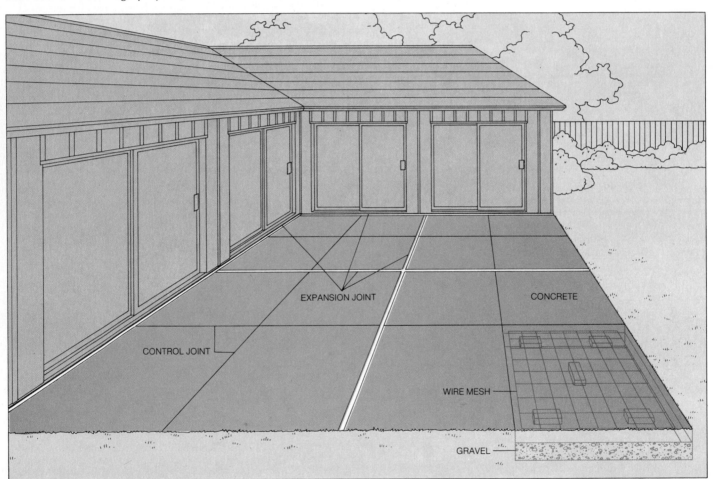

EXPANSION JOINT

CONCRETE

CONTROL JOINT

WIRE MESH

GRAVEL

Anatomy of a reinforced slab. A 6-inch gravel bed is topped by wire-mesh reinforcement to strengthen the 4 inches of concrete for a large slab such as this patio. Expansion joints are placed 8 to 10 feet apart in both directions in the slab, and between the slab and an abutting structure such as the house. Control joints to limit cracking are tooled halfway between the expan-

sion joints. A patio is sloped in one direction only, away from the house, so that water will run off it into the ground. How a driveway is sloped depends on the contours of land near the garage.

Making a Patio

1 **Grading a patio.** Lay out the patio, excavate the site and place support stakes, following the instructions for a sidewalk (*pages 42-45*). However, plan to place stakes inside form boards, and leave an extra ½ inch of space between the house and the stakes to accommodate the expansion joints. To grade the patio so that its slope will carry water away from the house, mark the stakes along the edge of the house at the desired level. Then at each stake 6 feet away, pencil a line 1 inch below the leveling marks. Another 6 feet away, drop the grade to 2 inches below the leveling marks, and so on. Attach form boards (*pages 47-48*) using the marks as guides for board tops.

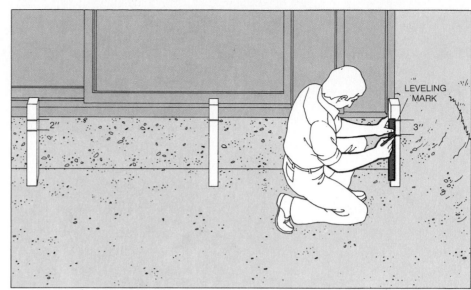

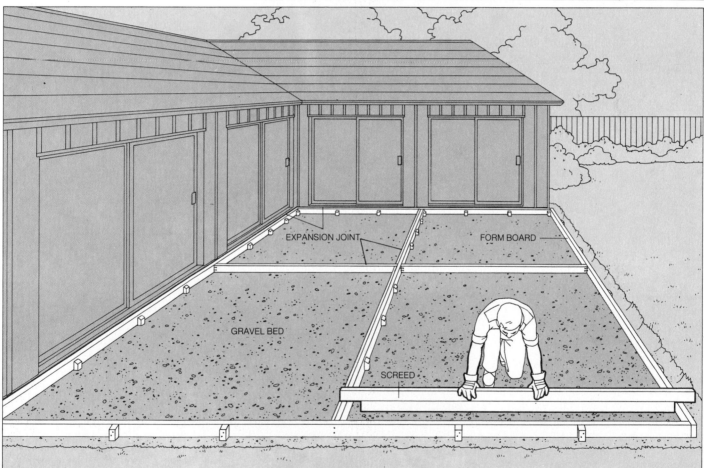

2 **Expansion joints and gravel bed.** Place expansion joint material against the house and build forms next to it, using 3-foot lengths of 2-by-4s, which are easier to pull out after the concrete has been poured. Divide the patio into sections no larger than 10 by 10 feet. Set up additional form boards running the length of the patio with expansion joints against these forms on the side that will be poured first and stakes on the other side. Lay a gravel bed 6 inches deep and screed one section at a time. Across the width of the patio, place expansion joint material, holding it in place with four 3½-inch nails driven part of the way into the form boards (*page 49*).

3 Laying the wire mesh. Place the cut end of the wire mesh 2 inches from the form board and weight it with a masonry block. Walk backwards, unrolling the wire mesh. Cut it off 2 inches from the form board at the other end of the patio section. Because the wire is stiff, the ends will remain curled. With a helper, turn the mesh over and flatten it by walking on it. Bind the ends of the wire together where two strips overlap. Just before pouring the concrete, pick up the wire mesh and support it 2 inches above the gravel bed with bricks or stones placed 2 to 3 feet apart.

4 Removing form boards and filling spaces. After striking off the poured concrete (*page 51*) in each section, build a bridge with a ladder supported by masonry blocks and topped with a plank. One at a time, pry up sections of form board with a crowbar. Lift them out by pulling upward and wiggling them from side to side to loosen the stakes. Leave the center boards temporarily in place. Shovel concrete into the resulting spaces, and finish it off with a darby. After pouring the adjacent section, extend the ladder to span both sections. Finally, pull the form boards from the center after striking off the adjacent sections.

5 Bull floating. Since you cannot reach to the middle of a large slab, use a tool called a bull float in place of a darby for finishing. You can make one using a 1-by-12 board 4 feet long with a 2-by-2 handle 12 feet long. Cut its end at about a 10° angle so that when screwed to the 1-by-12 board, the other end of the handle will be at eye level. Have a helper place the float at the opposite edge of the slab from where you stand. Keeping the float board flat against the concrete, draw it toward you. Repeat the operation from the other side of the slab.

A Patio Poured in Sections

Making permanent forms. With rot-resistant wood—pentachlorophenol-treated pine, redwood, locust or cedar—the forms can stay in place permanently, permitting you to divide a large patio into a grid of small sections to be poured a few at a time. This way one man can construct a patio at his own pace. Any rectangular pattern that pleases you will serve, but keep the sections no larger than 25 square feet, the amount a man can handle by himself in a day.

EXPANSION JOINT

Building Permanent Forms

1 **Laying out the grid.** Place the support stakes for perimeter forms inside the form boards. Except at the corners, drive them 1 inch below the boards so that the stakes will not show. After completing the perimeter forms, lay out the grid pattern with strings and stakes, maintaining the slope of the patio from house to outside edge. At each point where strings cross, drive an 8-inch, 2-by-4 stake, preservative-treated, into the gravel base. Then establish the line for the bottom of the form boards by running a taut string from underneath the perimeter form boards to intersect the 2-by-4 stakes. Using a sledge hammer, drive the stakes down to that level.

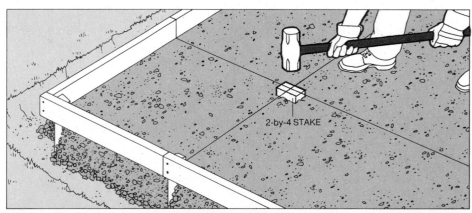

2-by-4 STAKE

2 **Securing the forms to stakes.** Rest one form board on a 2-by-4 stake and drive a 3-inch galvanized nail into it at an angle to fasten it to the stake. Butt another board against the first, and secure it with an angled nail. Then nail through the first board into the end of the second. Place wire mesh cut to size in each section of the grid.

3 **Nailing the sides of the forms.** To bind the concrete to the form boards, drive 3-inch galvanized nails into the sides of the boards about midway between top and bottom at 1-foot intervals. Hold a mallet or another hammer on the opposite side of the 2-by-4 to prevent the forms from moving, and leave about half the nail sticking out. Protect the tops of the boards with heavy-duty masking tape while pouring and finishing the concrete.

A Driveway that Lasts

A concrete driveway is more expensive initially than one of asphalt, but if properly reinforced and made thick enough for the traffic it will bear, it lasts many years with no maintenance. It is built like a patio, but may have to be thicker, requires heavier 6×6-6/6 reinforcement and is graded differently.

For a single-car garage the driveway should be at least 10 feet wide; for a two-car garage, a 21-foot apron is required, but can narrow to 10 feet. If only passenger cars will use the driveway, a 4-inch slab is sufficient. If heavy vehicles like oil-delivery trucks will drive over it, make the concrete 6 inches thick. Set expansion joint material against form boards at 8-foot intervals. Pour and strike off each 8-foot section separately; then remove the form board separating one section from the next and fill the space left with concrete as shown in the patio instructions *(page 60)*. Tool a control joint halfway between expansion joints.

Make a scale drawing of the driveway including existing features that affect the driveway's alignment, such as garage, plantings and house. Also plot the street access, and turning area if you need one. Make sure the length of the driveway will slope no more than 1¾ inches per foot; otherwise, car bottoms will scrape. Check local codes for breaking through curb or sidewalk.

How you slope the driveway sideways depends on the land around it. If the site is level and above the street, a crowned driveway, raised 1 inch in the center, is preferable. Water will run off both sides and toward the street. If you want water to drain to one side only, pitch the driveway in that direction like a patio. Where the garage is below street level, it is best to make a concave driveway. You must then install a drain where the driveway meets the apron of the garage. Check your local building department for drainage requirements.

Curved strike-off boards. Make a strike-off board of two boards, a 2-by-4 and a 2-by-6, the width of the driveway. For a crowned driveway, lay them edge to edge and nail a piece of scrap wood across the centers to hold them together. At the ends insert blocks of wood ¾ by ¾ by 1½ inches to bow the wood. Nail a scrap piece of wood near each end to lock in the bow. In order to make a driveway concave, fasten the ends of the two boards together first, place a block between the boards at the center and then fasten the center.

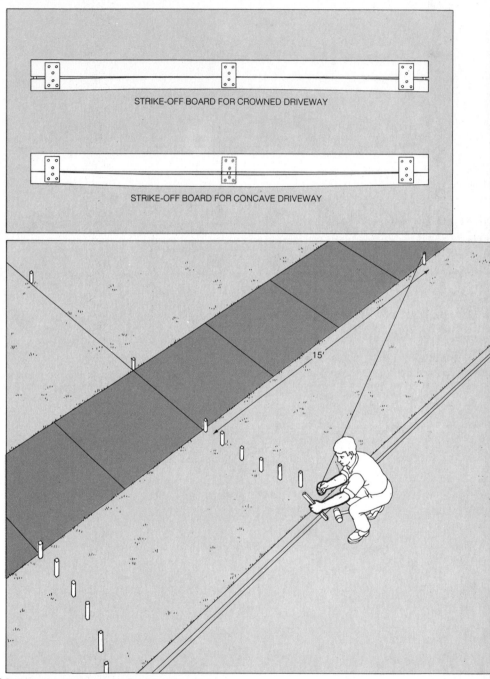

STRIKE-OFF BOARD FOR CROWNED DRIVEWAY

STRIKE-OFF BOARD FOR CONCAVE DRIVEWAY

15'

1 **Staking the street access.** Establish the curve of the driveway for the entry from the street by setting a stake at the sidewalk edge 15 feet from each side of the driveway. Drive a nail into the stake's top. Attach a 15-foot string to it to use as a radius to mark off the curve from driveway to street. Stake at 1-foot intervals.

2 Forms for the street access. Excavate to 10 inches below street level where the driveway will meet the street, and 10 inches below the level of the sidewalk. Grade the bottom of the excavation so it slopes up from the street to the sidewalk. Build straight wooden forms to the street. Then build curved forms to follow the curve plotted in Step 1, using sheet-metal garden edging cut to the height of the curbstone plus the height of the concrete—an additional 4 to 6 inches. Be sure the edging top is flush with the curbstone and slopes up to meet the sidewalk. Set expansion joints on the sidewalk sides, and between the driveway and the street.

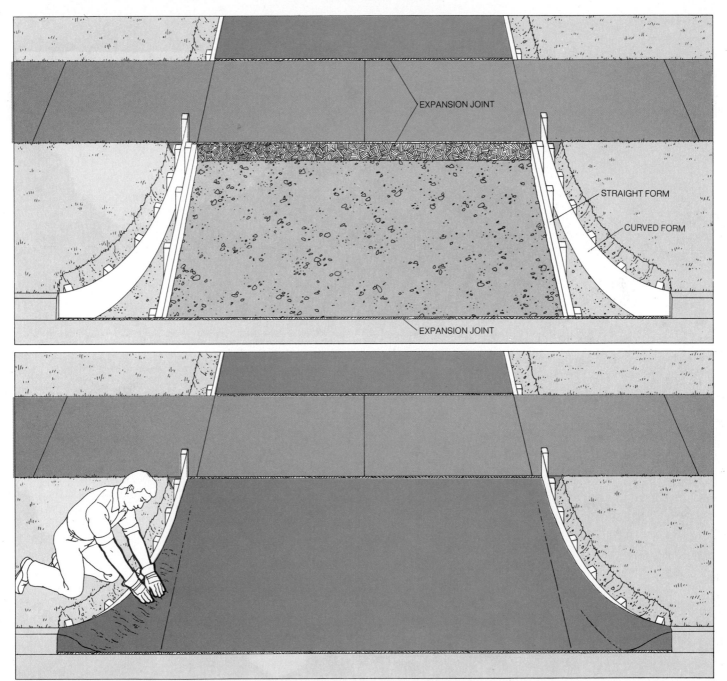

EXPANSION JOINT

STRAIGHT FORM

CURVED FORM

EXPANSION JOINT

3 Finishing the street access. After pouring the concrete, strike off the center of the driveway, pushing the excess concrete to the sides. Pull out the straight form boards and fill the remaining curved space. Finish the sides of the street access, shaping them with a darby or float. To blend the curb into the driveway, shape the concrete with your hands (be sure to wear gloves), tapering it gradually from the top of the curb into the driveway's curve and sloping it into the flat surface of the driveway. Finish this section with a small float.

Extending an Existing Slab

Sometimes a new concrete slab must be connected to an existing structure. You may want to enlarge a patio, for example, or carry a sidewalk around the house, or add a turnaround apron to the drive. There are two possible ways to handle such extensions, depending on the size of the planned addition. If it is large enough—about 20 square feet for a sidewalk, 40 square feet for a patio or a drive—the addition should be constructed just as a section of new slab would be (pages 49 and 59), with strips of expansion joint filler setting off the new slab from the old.

However, if the new area is relatively small—say, an extension of less than 5 feet to a sidewalk or of less than 4 feet to a patio, as shown here—then the new slab can be poured directly against the edge of the old. The two slabs are bound together with steel reinforcing rod at least ½ inch in diameter. The rod, available in masonry-supply stores, can be cut to length with a hacksaw.

Locking Together Old and New Concrete

1 **Installing reinforcing rod.** Excavate the area planned for the extension and clean the edge of the old slab, removing any dirt or damaged concrete. Using either a star drill or a hammer drill that has been fitted with a carbide-tipped bit ¼ inch larger than the diameter of the reinforcing rod you are using, drill 6-inch holes into the edge of the old slab. Position the first hole 3 inches in from the side of the slab and space the others 12 inches apart. Clear the holes of debris by squirting water into them from a hose. Poke mortar into the holes with a joint filler until they are halfway filled; then tap in 12-inch lengths of reinforcing rod.

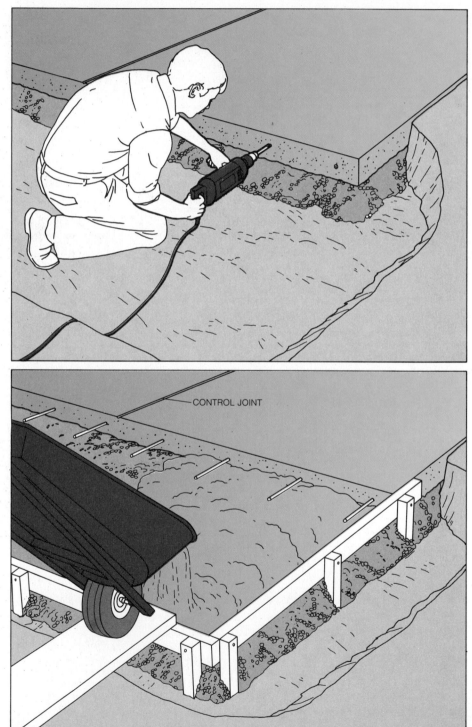

CONTROL JOINT

2 **Pouring the addition.** Construct forms as on page 42 and add a 6-inch bed of gravel. If the addition straddles a section of patio or driveway already intersected by an expansion joint, install a matching expansion joint in the forms for the extension, as described on page 49. Thoroughly wet the edges of the old slab; then pour the concrete. Finish the surface of the new slab as on pages 51-53, scoring it with control joints, as needed, to match the control joints in the old slab. Leave the forms in place for seven days.

Sturdy Footings for Strong Walls

Below the base of a wall or a building, usually concealed by earth or sod, lies a vital structural element—the footing. It is supported by solid ground or a bed of gravel. In turn, it supports all the weight of the structure above, which may be a 3-foot-high garden wall or a three-story-high home. Whatever its load, the footing takes the form of a long, narrow, flat-topped slab.

Footings designed for the light low brick and stone walls described in this book do not differ much from an ordinary concrete slab, poured just below the surface of the ground. On level ground with firm soil, the footings may not need wooden forms: a straight-sided trench with a gravel bed will serve for pouring concrete. More often, to make a square footing with a level top, you will have to build simple forms that are braced by crosswise cleats (right).

The footings for a building or for a freestanding wall more than 4 feet high are more complex, and they are strictly regulated by building codes. For a rectangular building plan, the footing corners must be true right angles. Because the heavy load such a footing supports will not float as a unit over frost heaves, all codes require that these footings be built with bases below the frost line—in some regions, several feet deep.

Codes generally prescribe dimensions, which are based on wall height and width, and they may specify drain pipes along the base. Reinforcement with steel rods is usually necessary. The basic techniques that are involved in constructing such strong load-bearing footings are illustrated overleaf. But because they are complex and must resist great stresses, these techniques are best attempted only after the fundamentals of building simpler footings, like the one shown at right, have been mastered.

Any footing, heavy or light, can be made with a fairly thin concrete mixture, because concrete, buried beneath the ground, is not subject to wear or weather. A typical formula calls for a 7-gallon mix for footing concrete—that is, a mixture containing 7 gallons of water for each bag of cement.

A Footing for a Low, Light Wall

1 Building forms. A garden wall less than 4 feet high can rest on a footing made like a long narrow slab that floats above the frost line; as a general rule, such a footing needs no reinforcing. Dig a trench that is about 2 feet wider than the footing, then lay in 6 inches of gravel, and build forms with stakes and form boards by the methods described for a concrete sidewalk (pages 45-48). Be sure to drive the stakes through the gravel bed and down into the earth beneath. The trench, gravel bed and forms must be absolutely level; do not follow a sloping grade or introduce a slope for drainage. Finally, using double-headed nails, sometimes called scaffolding nails, fasten 1-by-2 cleats called spreaders across the tops of the form boards every 3 or 4 feet.

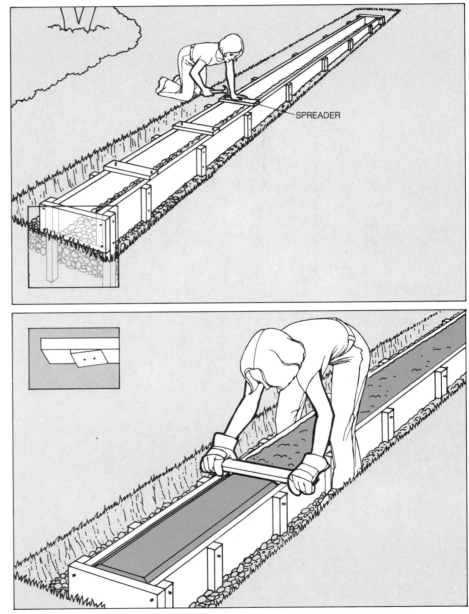

2 Setting the concrete. Mix and pour concrete for the footing as you would for an ordinary slab (pages 50-53), making sure that all the spaces immediately below the spreaders are completely filled. Remove the spreaders carefully without disturbing the semifluid concrete below them. Though the footing as a whole must be perfectly level, its edges should be slightly beveled or tapered to carry rain away from the base of the wall. Form the bevels with a home-made strike-off board (inset), consisting of a length of 2-by-4 fitted with wedges or shims about an inch wide and from ⅛ to ¼ inch thick. The wedges should fit snugly inside the form boards. Shape the bevels at the ends of the footing with a trowel, then finish and cure the concrete.

Footings for Buildings and High Walls

1 **Laying out.** To make the right-angle corners required for most buildings, use the 3-4-5 triangle method to line up footings. Stake one building line (AB) in relation to a property boundary or an existing structure. Along this line, exactly 3 feet from stake A, set stake C and drive a nail in it at the 3-foot mark directly under the AB line. Tie a 5-foot-long string to the nail in stake C, a 4-foot string to the nail in stake A. Where the ends of the two strings meet, set stake D. These three stakes should form a right triangle with sides of 3, 4 and 5 feet. Check the lengths with a steel tape (strings may stretch) and adjust stake D as needed. Extend the line from A through D to E, completing a second building line, and use the 3-4-5 method to lay out right triangles at B and E, completing the rectangular plan.

2 **Setting up batter boards.** At least 4 feet outside the corner stakes erect four batter boards. Each consists of two 1-by-6s or 1-by-8s nailed to three 2-by-2 stakes, and set in a right angle that encloses a corner stake. Transfer the strings of the building plan from the corner stakes to the batter boards, using a plumb line to make sure the strings intersect precisely over the corner stakes. Mark the string ends on the boards with a nail.

Remove strings and corner stakes so you can dig a footing trench—the plan can always be reconstructed by running strings from the marked points on the batter boards. Mark the batter boards to fix lines that lie outside the building plan: several inches outside for the edge of the footing, another foot or so for the edge of the footing trench. You can mark any of these lines directly on the ground by running strings between the batter boards and dribbling sand over the strings. The lines will show up as clear, straight, unsanded lengths along the ground.

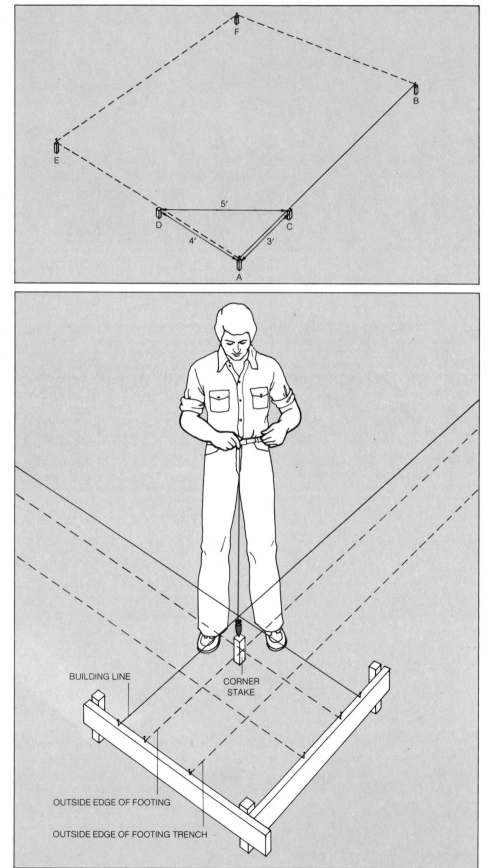

BUILDING LINE

CORNER STAKE

OUTSIDE EDGE OF FOOTING

OUTSIDE EDGE OF FOOTING TRENCH

3 **Laying in reinforcing rods.** Dig a trench for the footing and fill the bottom with a 6-inch layer of gravel or crushed rock. The top of this layer should be several inches below the frost line in your locality. Build forms for the footing and, before fastening spreaders across the tops of the forms, run at least three reinforcing rods from one end of the forms to the other. These steel rods come in 20-foot lengths and ⅜- or ½-inch thicknesses. Tie them with wire to rocks or bricks so that they lie about a third of the footing depth above the bottom. To splice rods, overlap them about 18 inches and fasten them with wire.

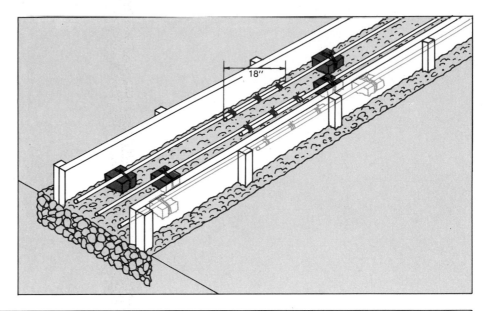

How Deep to Build Frost-Resistant Footings

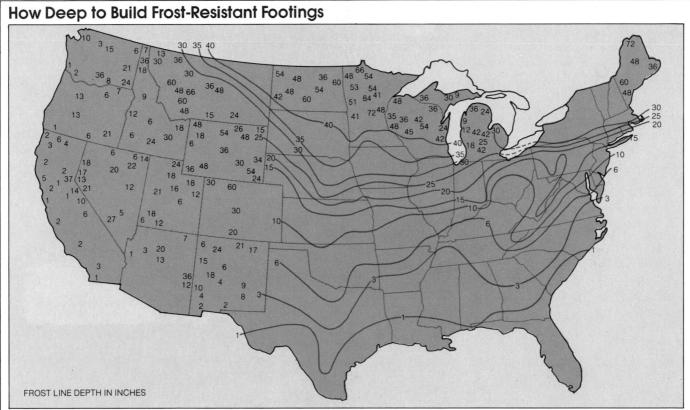

FROST LINE DEPTH IN INCHES

Every building and large, free-standing wall has an underground enemy—the water in the soil beneath it. In winter some of this water may freeze; it expands and moves the soil around it. In a thaw melting ice contracts and the soil shifts again. Such heaving can crack footings and damage a wall or building. To prevent such damage, a footing for heavy loads must rest on undisturbed soil below the frost line—the deepest penetration of frost.

As the U.S. Weather Bureau map *(above)* shows, the frost line varies from an inch in southern Texas to 6 feet in northern Maine. But the depth of the line is also influenced by soil composition, altitude and weather patterns. Thus, the map shows long lines of equal depth in the central and southern states; in the west and north, where high mountains and varied soils disrupt these even patterns, local depths may differ widely within an area.

Your building code may call for footing depths greater than those on the map, for reasons ranging from local custom to a known danger of earthquakes. But the required depth will never lie above the frost line.

Rugged Steps of Poured Concrete

When you need to replace worn-out wood stairs outdoors, or provide new ones for an added doorway, poured concrete is often the best choice: concrete steps are fairly simple to make and, once built, are all but indestructible. The method shown here, based on standard form-building and pouring techniques, is suitable for a structure no more than about 2½ feet high that is not attached to a building.

The only tricky part of building four or fewer steps is laying out the steps' profile on the plywood form sides: the design should not be a series of rectangular jogs but, for comfort and weather resistance, a very gentle saw-tooth, with the tops, or treads, sloping down and the vertical sections, or risers, sloping forward (bottom). For safety, it is crucial to get the dimensions right. The depth of the tread must be related to the height of the riser, and these two dimensions should total about 18 inches. Theoretically, an extremely high riser can be used with a shallow tread, or the other way

round, provided their sum meets the standard. As a practical matter, however, risers should be 6 to 8 inches high, and they should be combined with treads of 12 to 10 inches. You will have to determine the dimensions for your project by measuring and calculating (Step 2).

With the dimensions settled, you can draw plans and estimate materials. After you have figured out the required amount of concrete (page 40), you can mix what you need in a power mixer, but it is usually much simpler to order from a transit-mix company. You will also need five-ply plywood; 2-by-8s and 2-by-4s for forms and braces; double-headed nails; 1¼-inch-size gravel; form oil, which is spread on the forms to prevent them from sticking to concrete; and flexible mastic to make a joint, sealing the space between the steps and house. The gravel base is simple to make alone, but when the transit-mix truck arrives, have a friend help you move the concrete from the truck to the form in a wheelbarrow, and shovel it into place.

1 Finding the height. Level the ground approximately 6 feet in front of the doorframe. Then measure from the ground to the underside of the sill. This will be the rise, the vertical distance that the stairs will fill when completed.

2 Figuring dimensions. All steps should be at least 6 inches wider than the door opening. The topmost step, or platform, should be at least 3 feet deep to provide an area for entering and leaving safely. Divide the rise found in Step 1 by the number of steps you wish to build to find the height of each riser. Subtract the riser height from 18 inches to find the tread width. Thus, if the rise comes to 24 inches, you could make three steps, each with an 8-inch riser and 10-inch tread or, alternatively, four gentle steps, each with a 6-inch riser and 12-inch tread. From the dimensions you arrive at, make a plan and estimate materials.

3 Cutting side forms. Lay out the sides of the forms on a sheet of five-ply plywood using the dimensions selected in Step 2. Make the height of the forms 1 foot greater than the rise, since 1 foot of concrete and gravel will be below ground level. First draw risers and treads at right angles using a steel square. Slope each tread and the platform downward (dotted lines) to provide drainage, pitching them ¼ inch for each foot of depth. Slope the risers 15° forward (dotted lines).

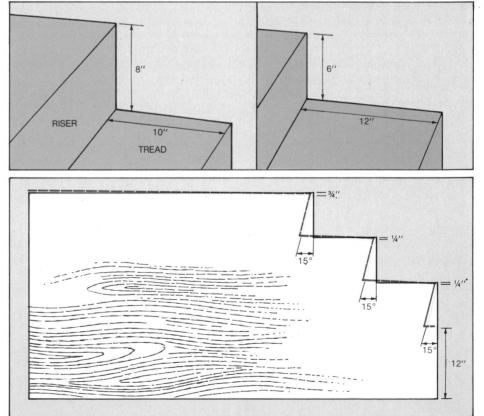

4 **Setting up the form sides.** Dig a hole 1 foot deep and about 1 inch larger all around than the steps. Tamp the ground firmly at the bottom of the hole. Nail vertical 2-by-4 braces to the outsides of the form sides. Place the form sides against the sides of the hole, ½ inch away from the house wall, using a steel square to make sure they are at right angles to the house, and a level to check that the edges are truly horizontal and vertical. Drive 2-by-4 stakes at least 8 inches into the ground about 18 inches away from the hole. Nail 2-by-4 braces between the form sides and the stakes. Pour in 6 inches of gravel and tamp.

5 **Completing the form.** Cut pieces of 2-by-8 to fit the risers, beveling the bottom edges so that your trowel will be able to reach and smooth the entire surface of the tread after the concrete is poured. Nail the riser boards to the form sides with double-headed nails. When the form is completely in place, apply a ½-inch layer of mastic to the house foundation (drawing) for an expansion joint. Coat inside surfaces of the forms with form oil to prevent concrete from sticking to them.

MASTIC EXPANSION JOINT

6 **Pouring concrete into the form.** If the mixer or delivery truck can be brought close to the job, pour concrete directly into the form through a trough; otherwise you will have to use a wheelbarrow. Begin by pouring concrete into the lowest step; when it is full, pour or shovel concrete into the second step. If the riser boards bulge outward, brace them with 2-by-4s (inset).

Pour enough concrete to overfill the form slightly. Drive shovels into every corner to make sure the form is completely filled and air pockets are eliminated. When pouring is completed, plunge shovels down into the concrete along the inside edges of the form to force large aggregate away from the outside edge of the concrete.

Blend and smooth the freshly poured concrete as shown on pages 50-53, finishing the surfaces after the bleed water has evaporated, and then curing the structure for a week. Finally, when the steps are cured, you can remove the forms.

Creating Free-Form Shapes: A Bowl for a Pool

Pouring concrete in the shape of a bowl is even simpler than pouring it for a slab, and the technique involved makes a variety of useful, decorative structures. A water-filled bowl, sunk flush to the surface of a lawn or garden, serves as a wading pool or a picturesque setting for statuary. A recirculating pump turns the bowl into a fountain. With a runoff channel in the rim of the bowl, it becomes a fishpond (below). Any of these things can be built without wood forms. Generally, the concrete can be poured directly onto tamped earth without a gravel bed. Since a bowl need not bear much weight, its concrete sides and base need be no thicker than 4 inches.

Small concrete bowls for any purpose can be constructed like the basic 6-by-4-by-2-foot fishpond shown on these pages. It requires just under ³/₅ cubic yard of concrete, is small enough to fit into a backyard and is so shallow that its 350 gallons of water are self-aerating. Because a bowl is built without forms, it calls for a stiffer concrete mixture than a slab: a 6- to 6½-bag mix with minimum slump (page 41). If a truck can get close enough to the excavation site to reach it with chutes, you can trowel and finish the concrete yourself in less than half a day with ready-mix; if not, you will probably need an assistant.

In other phases of construction you will use some tools and techniques that differ from those designed for slabs. A standard square or rectangular tamp, for instance, is too angular for the bowl's rounded contours—but a spare tire turns out to be perfect for tamping. A length of plastic around the top of a shaped excavation keeps dirt from mixing with the concrete. The same wire reinforcing mesh used in a slab is also used in the bowl, but it must be bent into the proper shape first. And a bowl, unlike a slab, must be covered during the curing process with a tarpaulin, a piece of burlap or an old double-bed sheet to absorb surface water, which would otherwise flow to the bottom of the excavation.

If the finished bowl is to contain water, it must be sealed. For a wading pool, any waterproof paint or silicone sealer will do. If you intend to use the bowl as a fishpond, buy a neutralizing agent in powder form from a garden store or concrete dealer. When the powder is mixed with water, it becomes a paint that both seals the pond and neutralizes the alkalis in the freshly cured concrete.

Stock a fishpond or water-plant pool carefully to prevent it from becoming a mosquito-breeding nuisance. Goldfish are a surprisingly effective form of mosquito control; in addition, consider small catfish, tadpoles and Japanese scavenger snails. Use weighted pots or boxes on the pond floor for oxygenating plants like loosestrife or tape grass and for purely decorative water lilies. Water hyacinths and duckweed will float, roots and all, on the surface.

The Anatomy of a Concrete Bowl

A pond made without forms. The excavation for a bowl like this fishpond must be 4 inches larger in every dimension than the inner concrete wall and floor of the finished bowl. Because the concrete is poured without forms, the sides of the hole must slope at an angle of no more than 60° or 70°—a steeper slope would cause the concrete, while it is being poured, to slide from the sides of the bowl to the bottom. Wire mesh (inset cross section) reinforces the concrete. For a fishpond, a runoff channel in the rim permits excess water to flow away at a single point.

Preparing the Excavation

1 Tamping the earth. Outline the bowl with rope or a length of garden hose—for the fishpond shown in this demonstration, the outline should be roughly 6 by 4 feet with an extra allowance of 4 inches on all sides for the concrete—and mark the outline on the ground with a spade. Dig the excavation from the edges down until you reach a uniform bottom depth—for this fishpond, about 28 inches. Tamp and compact the earth by bouncing a spare tire vigorously around the bottom and along the sides of the excavation.

2 Laying in the reinforcing mesh. Bend a single piece of 6-by-6-inch wire reinforcing mesh to fit inside the entire bowl shape of the excavation, cutting into the edges of the mesh to help mold it into the bowl shape. A 12-foot-long piece of 5- or 6-foot-wide mesh should fit the basic fishpond but the fit does not have to be exact—the wire mesh need not reach all the way to the top. After shaping the mesh, remove it and run a continuous strip of black plastic sheeting around the top of the excavation, with the inner edge extending about a foot down the sloped sides. Line the bottom and sides of the hole with 2-inch-high rocks or pieces of brick spaced at 1-foot intervals; replace the mesh on these supports.

3 **Marking the depth of the concrete.** The concrete lining of the bowl must be 4 inches deep. To control the depth, drive 10-inch-long, 2-by-2-inch stakes at 1-foot intervals into the bottom and sides between the openings in the mesh. Mark a line 4 inches from the ground on each stake.

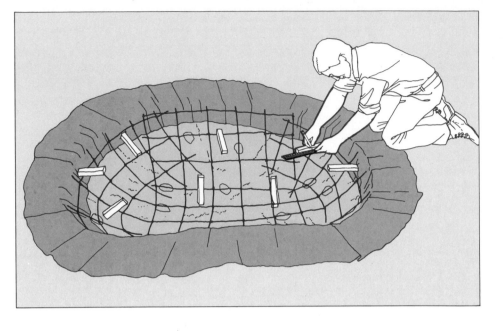

Shaping the Bowl

1 **Pouring and smoothing the concrete.** Pour the concrete into the center of the excavation, then lay a 2-by-12 board across opposite sides of the hole, propping the ends up on cinder blocks. Using the board as necessary for easy access to every part of the excavation, work the concrete outward and up the sides with the back of a square shovel. If the mesh sinks below the 2-inch level of brick or rock supports at any point, lift the mesh with a rake. As you bring the concrete level up to the 4-inch marks on the stakes, pull the stakes out and fill the holes they leave with concrete. Shovel all excess concrete onto the plastic strip around the rim of the excavation. Finally, smooth the concrete with a wood float. Caution: Float the sides lightly; excess surface water from vigorous floating will trickle to the bottom, causing hairline cracks.

2 Forming the lip. The simplest tool for edging the concrete around a small bowl is an ordinary 2-pound coffee can with the ends removed and the body cut in half lengthwise. Using this half cylinder as a portable mold, shape the concrete over the plastic rim, smoothing irregularities with a pointed trowel as you go.

3 Making a runoff channel. Immediately after forming the lip, cut a 3-inch-wide semicircular section out of it and embed into the opening 1-by-1-inch wire screening, bent double to avoid sharp points. The channel makes an overflow outlet in a downpour; the wire screen prevents fish from swimming out with the overflow.

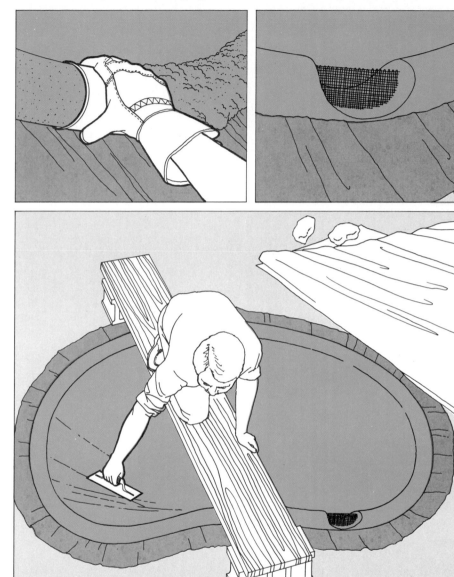

4 Troweling and curing. Cut off the black plastic strip flush to the outside of the lip with scissors, and cover the entire pond, board and all, with a tarpaulin, burlap or double sheet, weighted at the edges with rocks. The covering will keep hairline cracks from developing in the concrete between three separate trowelings. Wait 30 minutes to an hour, until all water sheen on the surface disappears, then go over the surface of the concrete with a square trowel. Cover the pond loosely again and wait about 45 minutes —at this point the concrete should be barely spongy to the touch and the weight of your hand should leave a faint impression—and trowel a second time. Cover the pond again for 45 minutes. On the third troweling your hand should leave no impression in the concrete and the trowel should make a ringing noise as you work. Wait 10 or 15 minutes for any residual water to evaporate, then remove the work board. Wet the fabric covering and lay it directly on the concrete inside the bowl and up over the lip. Keep the covering damp at least three days, or until the concrete is cured.

How to Empty a Pond with a Siphon

Even in a moderately cool climate, a 2-foot-deep fishpond can safely be left full year-round. If ice forms at the surface during the winter, it normally acts as an insulating skin that keeps the water below from freezing. However, in extremely cold northern climates or during a severe cold snap in moderate climates, make sure the ice is not freezing solid. Fish can live in a dormant state under a surface layer of ice but they cannot survive a solid freeze; you will have to take them inside, along with your plants for the duration of the winter or cold snap.

Regardless of the weather, you may want to drain the pond periodically for painting or cleaning. There are several ways to do the job, from bailing the water out with a bucket to pumping it out with a submersible power pump. But if your property has ground lower than the pond, you can siphon the water out with a garden hose.

The most efficient way to siphon with a hose is to attach it to a tap and run water through to force the air out. Then close the nozzle end tight and set it on the ground below the pond. Turn the tap on again and fill the hose completely. Finally, turn the tap off and unscrew the hose, cupping your hand over the end to hold the water in. Submerge the tap end of the hose in the pond and open the nozzle end: the water should immediately start to drain from the pond.

If your garden tap is inconveniently far from the pond, use this alternate, somewhat slower method: Submerge the hose in the water with both ends open. When the hose is full (be patient—it will fill slowly, and it must be entirely full, with no air bubbles), cap one end with your hand and carry it to lower ground. Uncap the hose to let the pond drain.

3

A Cornucopia of Blocks

Shapes and textures for every need. The sampling at left suggests the effects—elegant or rustic, naturalistic or geometric—that can be achieved with easy-to-handle blocks. Clockwise from top center are: paving bricks on a standard concrete block, quarried rubble on a half-sized concrete block, a quarry tile on slate flagging, a decorative screen block and a fieldstone.

For many, the pleasures of building with blocks reach a peak in the kindergarten years and fade away soon afterward. For a lucky few, however, these delights last a lifetime, growing deeper and more complex with experience and skill. Such individuals find a compelling attraction in working with masonry blocks: bricks, stones, tiles and concrete blocks. The planning of block paving or walls, the agreeably repetitive yet constantly changing task of fitting and mortaring the individual pieces, the sense of achievement at the completion of the job—these are the rewards of masonry as a hobby, and they increase with the years.

Curiously, the appeal of the mason's work has been especially notable among men known for their literary talents (as well as other things). Within the present century, critic H. L. Mencken and poet T. S. Eliot have both been enthusiastic amateur bricklayers. But the most famous of these part-time masons was Winston Churchill. At Chartwell, his country estate, Churchill gave much of his spare time to projects that ranged from the laying of brick paths and garden walls to the building of small cottages. Whatever the brickwork task, he found satisfaction in it, for it brought rest and strength to what he called "the tired places of his mind."

Such amateurs of masonry pursue their hobby for both esthetic and practical ends. On the esthetic side, there is the contrast of individual blocks on the one hand, and blocks massed in patterns on the other. With masonry blocks, the total effect becomes greater than the sum of the parts. Brick, concrete block, stone and tile create eye-catching patterns, made more dramatic by sunlight and shadow, that a monolithic material like smooth poured concrete ordinarily cannot be expected to produce.

Each category of block brings a special character to backyard projects. Bricks give ruggedness and pleasing symmetry; concrete blocks give an impression of strength and massiveness, even though many of them are shaped like fretwork *(left)* to form decorative screens; stones give an artless-seeming rusticity; tiles can give elegance and smoothness, and there is enough variety in the choice of colors to inspire complex designs.

From a practical standpoint, blocks are easy to use. Unlike concrete, which presents a mass of material that sometimes must be dealt with in minutes and may also require a platoon of assistants, blocks can be installed over a period of several weeks and their assembly can be a one-man (or one-woman) project. Even the children can participate in a job that uses bricks and tiles. The builder is able to set his own tempo, proceed just as far as he wants and quit when he decides to.

Modern Bricks: 10,000 Shapes, Sizes, Textures

Bricks have been used for at least 5,000 years and they are still the most popular masonry blocks for all sorts of structures, from walls, steps and fireplaces to walks and patios. Bricks are compact and light to handle yet also strong and durable. Their beauty seems to improve with age —a quality that has inspired some manufacturers to produce new ones that simulate the chipped edges, paint splatters and patina of old ones.

Modern bricks are made in over 10,000 combinations of shape, size, color and texture, but within this staggering array you will find that most bricks fit into three categories: building, face and paving bricks.

Building bricks are an all-purpose, economically priced type suitable for any outdoor use. Generally red in color, most of them are 7½ to 8 inches long by 2¼ inches deep and 3¼ to 3½ inches wide. Face bricks—which come in a great range of sizes, in colors from almost white to purplish-black, as well as in textures from rough (opposite) to glassy-smooth —make distinctive walls, steps and even barbecues, but most are too textured (and expensive) for walks or patios. There you will want to use paving bricks, which are smooth-surfaced and may be shallow (sometimes only ½ inch thick so that they can be laid, like tiles, in a mortar bed). Paving bricks come in assorted colors, mostly shades of red, and in over 40 sizes and shapes, including squares and hexagons.

In choosing bricks for all outdoor masonry projects, weather-resistance must be considered. All three types are available in compositions whose resistance to frost is indicated in the manufacturer's specification sheet by the code SW, MW or NW. Bricks given an SW rating can withstand severe weathering, MW moderate, and NW no weathering. Only SW-rated bricks can safely tolerate direct contact with the ground, so do all paving with SW paving or SW building bricks. SW building or SW face bricks may also be used for walls, steps or barbecues in cold regions.

At most dealers, bricks are priced by the thousand, but are least costly to buy in prepackaged cubes of 500 to 1,000 bricks. Buying by the cube is not only economical but also ensures that all your bricks are approximately the same color and size. Small irregularities, however, are inevitable.

To estimate how many bricks you will need, first calculate in square feet the total area that needs to be covered. If the area is irregular, divide it into squares, rectangles or circles so that you can use standard arithmetical formulas for each segment, and add up the segments. If you need a double layer of bricks—as in a wall—do not forget to allow for both surface areas.

Calculate the number of bricks you will require to cover a square foot as follows. For unmortared paving, multiply the length by width (in inches) of the brick surface that will be visible; divide the result into 144. For mortared paving, walls and brick-sheathed steps, add ½ inch to each dimension before multiplying them, then divide the result into 144.

Next compute the brick quantity by multiplying the number of square feet in the area to be covered by the number of mortarless, or mortared, bricks to a square foot. Finally, add at least 5 per cent for cutting and breakage.

To estimate mortar requirements for paving, figure 8 cubic feet for each 100 square feet of bricks. For a double-thick wall, figure about 20 cubic feet of mortar for each 100 square feet of bricks, and for brick-veneered steps, about 12 cubic feet for each 100 square feet. A ½-inch-thick mortar bed for paving, walls or steps will require about 5 cubic feet of mortar for each 100 square feet.

Before ordering bricks, clear a generous delivery space close to the street or driveway and as near as possible to the site where you plan to work. Build there a pallet or wood platform of boards laid across 2-by-4s. After the load is delivered, cover it with plastic sheeting.

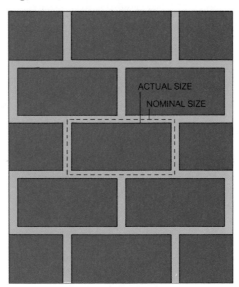

ACTUAL SIZE

NOMINAL SIZE

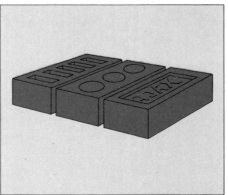

Brick sizes. Brick dealers and masons sometimes refer to a brick by its nominal rather than its actual size (dotted lines), which is the size as measured after it has been mortared into a wall with ⅜ or ½ inch for the mortar joint included. However, bricks are also specified in actual sizes. Either measurement represents only an average, in any case, because bricks from the same lot may vary in size as much as 3 per cent.

Solid and cored bricks. Some bricks are solid blocks, flat on all surfaces or with a slight depression and the manufacturer's imprint on one side. However, some have rectangular or round holes called cores, which run through the bricks from top to bottom (two bricks at left). Cored bricks produce stronger walls because some of the mortar will trickle down into the holes; the tops of walls made with them must, of course, be covered with solid bricks. Paving bricks do not have cores, but you can use cored building bricks instead for paths and patios wherever the pattern you have selected calls for bricks to be set on their sides or ends.

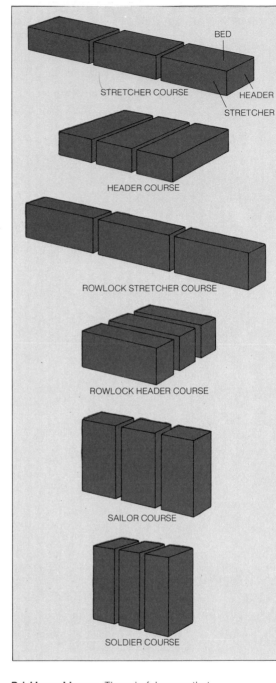

STRETCHER COURSE

BED
HEADER
STRETCHER

HEADER COURSE

ROWLOCK STRETCHER COURSE

ROWLOCK HEADER COURSE

SAILOR COURSE

SOLDIER COURSE

Bricklayers' jargon. The colorful names that masons apply to positions of bricks and arrangements of rows come from terms for brick surfaces: the long sides are called stretchers, the ends headers, the tops and bottoms beds. When bricks are laid on beds with headers abutting, the stretchers are exposed and the row is a stretcher course. When they are laid on beds with stretchers abutting, the headers are exposed and the row is a header course. Bricks laid on stretchers form a rowlock stretcher course when headers abut, a rowlock header course when beds abut. Bricks laid on headers form a sailor course when stretchers abut, a soldier course when beds abut.

The Popular Types

Common building brick. This gives walls a rustic look and paving a practical surface.

Smooth brick. Untextured on one side and both ends, it is useful for paving and walls.

Stippled brick. Roughly mottled on one side and both ends, this makes distinctive walls, but may be too uneven for paving.

Rug brick. Striped with deep grooves on one side and the ends, rugs form handsome walls.

Sand-finished brick. Grit, impregnated on one side and both ends, makes this texture suitable for nonslip paving and is attractive in walls.

Water-struck brick. Textured on both sides as well as the ends, the water-struck type looks handmade and is used in walls.

Laying Bricks to Walk On

Bricks will produce an attractive and rugged path or patio if you simply lay them directly into the ground. Traffic, however, will shift the bricks out of place, and frosts will heave them up and break them. For more permanent paving, your best bet is to set the bricks into a bed of sand, or to lay them in mortar over a concrete foundation.

The chief advantages of brick paving are its handsome appearance and ease of construction—a child can help. However, unless the crevices between bricks are tightly mortared, they sprout weeds and catch toes and heels; brick paving is also prone to icing.

The six classic patterns for laying bricks are shown at right. All the variations shown on the opposite page are based on these six patterns. An identical number of bricks is required to pave the same area with any of the classic patterns or the variations. Some patterns require cutting bricks into segments and when you stand bricks on edge rather than lay them flat you will need half again as many to pave a prescribed area.

A plain pattern such as jack-on-jack will serve well for small spaces since the design becomes clear after the first few rows are set out. More intricate patterns such as herringbone and basket weave are better when you are paving a larger area since several pattern repeats are required before the design reveals itself clearly. The bricks required for all but the jack-on-jack pattern must be approximately twice as long in actual length as they are wide.

To make sure your choice of patterns will work out as imagined, draw your path or patio area to scale before you even order bricks. Cut out miniature paper or cardboard rectangles of the size of your bricks and experiment with different combinations on your scale drawing.

The Basic Paving Patterns

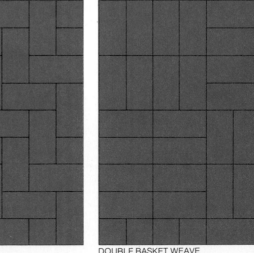

JACK-ON-JACK

BASKET WEAVE

RUNNING BOND

HALF BASKET WEAVE

HERRINGBONE

DOUBLE BASKET WEAVE

Some Paving Variations

Shifting a pattern diagonally. Arrange your pattern repeats to fit diagonally, rather than perpendicularly, against the outside edges of your path or patio. Cut some of the bricks at an angle to fill the ends of each course.

Standing bricks on edge. By turning all of the bricks in any pattern on their sides, you get a lighter, more delicate-looking effect than when the bricks are laid flat. However, you will also need about 50 per cent more bricks to cover the area. To use bricks edgewise in a basket-weave or herringbone pattern, select bricks whose thickness is equal to approximately one third of their length—2⅔ by 8 inches, for example.

Shifting the joints. A variation on running bond places the joints of the bricks in even rows at the one third or one quarter point of the bricks in odd rows, rather than at the conventional one half mark. You can also use two or three different fractional variations for successive rows.

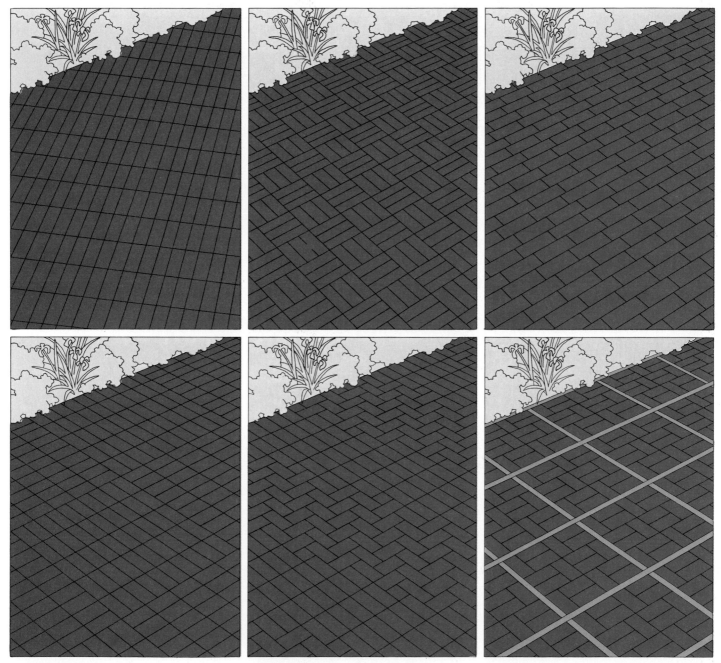

Changing directions. When two units of jack-on-jack are set at right angles to each other, a simple pattern becomes more dramatic.

Combining patterns. For large areas, you can vary the paving by mixing together two or even three patterns. In the drawing above, a fancy herringbone alternates with a plain jack-on-jack; if you try to combine two fancy patterns, one may detract from the other.

Gridding patterns. For structural strength, build wood grids into the path or patio. The patterns within the grids may be identical, as shown here, or may differ from grid to grid.

Classic Paths to Modern Paving

The methods for covering level surfaces with brick have not changed much over the last five thousand years—though they have been put to some new uses. The technique and tools used by American colonial masons to lay brick sidewalks and courtyards, for example, are now called upon for flower-bed borders and swimming-pool terraces. But the most common use of brick paving is still the construction of paths and patios with brick—whether mortared or unmortared. The brick in these basic instructions is paving brick, roughly twice as long as it is wide and about 2½ inches thick.

Unmortared brick is easy to work with, has a rustic charm and can be salvaged for reuse with a minimum of effort. In frost-free regions and in firm soil, the bricks can be laid directly on the ground, but in harsher climates such bricks may have to be releveled after a few years. For a more stable and long-lasting paving, lay unmortared bricks in a bed of coarse sand; simply enclose the sand with an edging.

In the drawings on the opposite page, four popular edging styles are shown being set in before the paving bricks have been laid. Three of the four edgings are brick, like the paving. Edgings may also be made of other materials, such as concrete. For decorative effect, attractive patterns can be achieved by wood edgings combined with wooden dividers that are set directly upon the bottom of the paving bed.

Permanently frost-resistant brick paving should be laid on concrete slab. The concrete foundation may be old—an existing sidewalk will do—or newly poured, following the instructions for constructing lightweight slabs. In either case, the job is best accomplished in two distinct steps: bond the bricks to the slab with regular mortar; then fill the spaces between bricks with thinner grout. Resist the temptation to try the professional mason's trick of doing the whole job with a single mixture. Without his skill in buttering each brick with exactly the right thickness of mortar, measured by eye, you are likely to wind up with an undulating, ill-fitting mess.

Paving without Mortar

1 Laying the dry run. Measure and stake the area to be paved as you would for a concrete slab, then estimate the number of bricks by the method used for unmortared paving *(page 76)*. To establish the final spacing of the bricks and to dig the paving bed and edging trench to exact, predetermined dimensions, you must lay out some bricks in a preliminary dry run. Enclose the entire area with edging bricks set on the ground, and lay out paving bricks within the edging. For a simple pattern like a running bond, you can save time by laying out rows of bricks along the sides of the area, leaving the middle bare, but you must be sure to complete the sides.

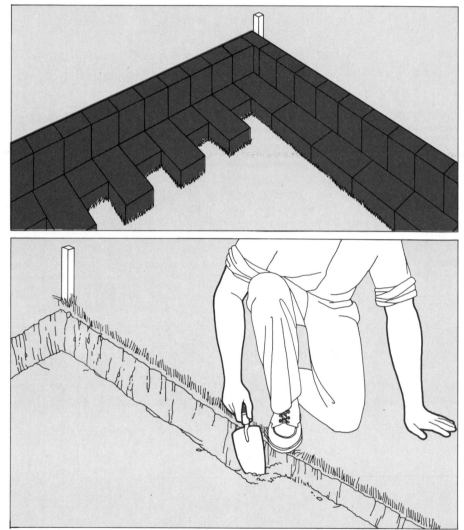

2 Edging the bed. Remove the bricks that you set out for the dry run and dig a 3½-inch-deep bed, following the slope for the existing grade and keeping the sides of the bed as nearly vertical as possible. Then, using a gardener's trowel, dig an edging trench 2¼ inches wide and 4½ inches deep along the inner borders of the bed. Set the edging bricks into the trench, tamping loose dirt along the inner sides of the trench to hold the bricks securely in place. They should form an upright wall that encloses the paving bed and rises to the level of the grade.

Some Edging Variations

A line of sailors. The simplest of all brick edgings is a straight line of sailors—bricks set on end with the long narrow stretcher sides abutting. A line of soldiers, with the wide bed sides abutting, makes a sharper and more attractive contrast with the paving bricks—but uses up almost twice as many bricks.

A gentle curve. Curved brick for a curved edging is available, but expensive. You can get the same effect with common bricks by angling sailors to form a gentle curve and filling the narrow wedges between the bricks with soil.

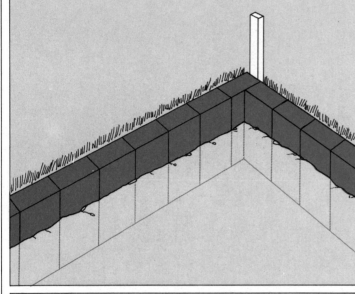

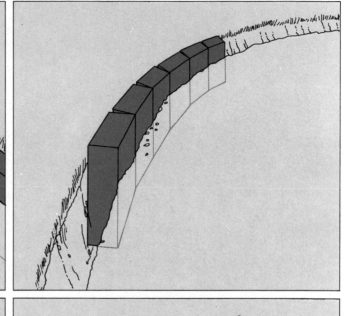

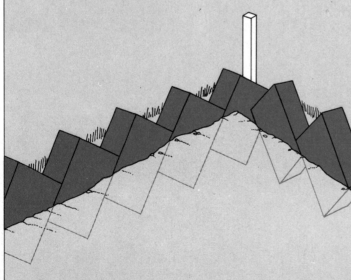

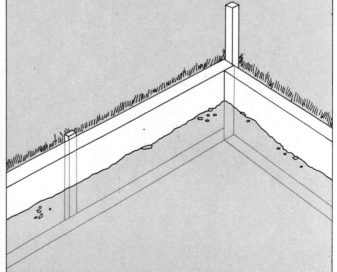

A sawtooth edging. Sailors tilted at an angle of 45° create a provocative optical illusion: when the paving bricks are in place, the edging seems to consist of a row of triangular bricks neatly cut to size. To enhance the illusion, dig a slightly shallower bed and trench, so that the paving and edging bricks project above grade. Caution: Lay this edging with care—the brick bases must be supported by packed earth and the tops leveled.

A wood edging. The forms used for pouring concrete adapt readily to the bed-and-trench construction of brick paving. Conceal unattractive pine or plywood edgings by setting their tops and supporting stakes at or slightly below grade. Show off a more attractive wood, such as redwood or cedar, by letting ¼ inch or so of the edging and paving bricks project above grade.

3 Making the sand bed. Tamp the earth of the bed, lay in a bed of coarse bank sand—the type used in mixing concrete—and screed the sand. The sand bed must be set just far enough below the edging so that bricks laid on it will be level with the edging. In a patio bed, set parallel 1-inch-thick wood strips on the earth 3 or 4 feet apart, pour sand between them and work a third across their tops to smooth the sand; then remove the strips and fill in depressions. To screed a bed for a path, use the method described for screeding gravel for a concrete sidewalk (page 49).

WOOD STRIP
1"

SCREED

4 Laying the bricks. Work the first two paving bricks into a corner of the bed. With the first bricks in place, use a length of thin twine wrapped around two bricks to set the alignment for the entire course. Place the two bricks outside the edging with the twine flush to the inner sides of the bricks you have laid. Complete the first course, using a level to check for a perfectly flat surface from brick to brick. Tap bricks lightly with a trowel handle to level them, and if necessary, add or remove sand under individual bricks. Lay bricks close together. The final course should be within ½ inch of the edging.

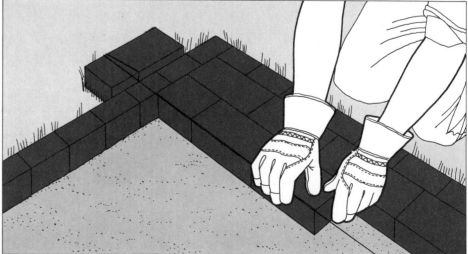

5 **Sanding the cracks.** With all the bricks laid and leveled, pour on the surface a bucket of pre-screened fine sand, which comes in bags (you can save money by sifting coarse sand yourself). Spread the sand back and forth with your hand or a brush or broom to fill the cracks.

SIFTED SAND

COARSE SAND

6 **Sweeping the sand.** After all cracks have been filled with sand, gently sweep excess sand off, working at a diagonal to the cracks—broom strokes parallel to the cracks can brush out the top layer of sand. Sanding may have to be repeated after the first application has settled.

Paving with Mortar

1 **Edging a concrete slab.** When covering an existing slab, be sure it is level and in good repair. If you pour a new slab, make one 4 inches thick, following the directions for lightweight slabs. You need not edge a slab lying so far below grade that the tops of the paving bricks will be at or near ground level. Otherwise, use edging to conceal the concrete and protect the outermost bricks from moisture and nicks.

Dig a trench around the slab, about 2½ inches wide and 5 inches deep. Hose down the edging bricks thoroughly and set them into the trench as sailors, flat sides facing out, using a forefinger or a ½-inch-thick scrap of wood as a spacer to allow for grouting between the bricks. Tamp dirt against the outside edge of the edging bricks to pin them to the concrete slab, and use a steel rule and a level to make certain that the bricks rise above the slab to the combined thickness of the paving bricks and a ½-inch bed joint.

2 **Preparing the bed joints.** Lay out paving bricks in a dry run to determine their fit along the width and length of the slab, leaving ½-inch spaces for mortar joints on all four sides of each brick. If necessary, reduce or enlarge the spacing to improve the fit. You can leave portions of the dry run in place as you work, serving as a guide to the placement of the mortared bricks.

Mix mortar in batches of 2 cubic feet, to cover about 6 square feet of the slab with a ½-inch-thick bed joint. If the total area to be paved is less than 12 square feet, you can mix extra mortar for the entire bed joint. Otherwise, observe the 2-cubic-foot limit, which yields no more mortar than you can easily cover in an hour, when it will begin to harden. Screed the mortar, then use a notched trowel to smooth and score it.

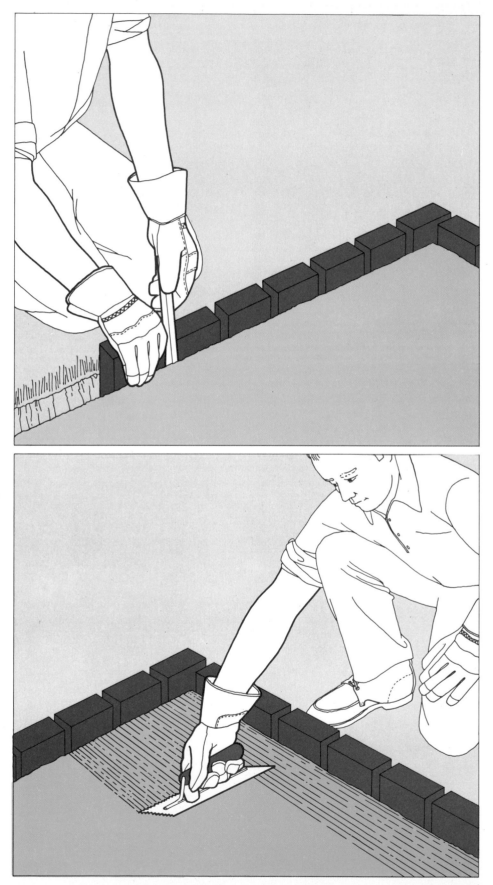

3 Laying the bricks. Soak the bricks thoroughly, then set them in place on the bed joint, smooth face up, using a stretched length of twine as a guide. On a small area, lay one complete course across the slab before starting the next; on a large one, lay rectangular segments, each about 2 by 3 feet. Position individual bricks by pushing them into the bed joint and tapping them lightly with the trowel handle; use a level to check for a flat surface from brick to brick.

4 Grouting the joints. On the day—or the weekend—after you lay the paving bricks on the mortar bed, mix a batch of thin mortar for grouting between the bricks, using the formula on page 76 to determine the quantity. The mixture should have the consistency of sour cream. Test it by dropping it from a trowel: if the mortar sticks to the trowel, or falls off in a single lump, thin it with water; if it runs off quickly in small drops, thicken it slightly with cement.

Hose down the bricks and lay in long, low ridges of grout directly on top of the joints. Work the grout into the joints with a small trowel or a brick jointer immediately after you throw each trowelful. Use the tip of the trowel or jointer to tamp the mortar to the bottom of the joint. When the

mortar nears the top of the bricks, use the full width of the tool to press down on the joints and fill gaps. Mortar until the joint overflows slightly. Use a trowel to shift or remove excess mortar, but do not smooth grouted joints at this stage.

About an hour after grouting—but before the mortar has fully hardened—remove most of the excess mortar from the joints with the edge of the jointer or trowel. Wait another three hours or so, then rub the joints smooth, using a stiff brush or a small sand-filled burlap bag. Brush excess mortar from the bricks and hose thoroughly. About two days later, when the mortar has completely set, remove any specks of dried mortar with a solution of muriatic acid, following the method for cleaning brick surfaces (*page 37*).

The Basics of Building with Blocks: A Brick Wall

Not everyone can match the expertise of Winston Churchill, an amateur mason who could lay a brick a minute (the average beginner's rate is about 75 to 100 a day) and was invited to join the bricklayers' union. But anyone who wants to can lay bricks well, in structures that are both useful and good-looking. The work is not difficult—bricks are light and easy to handle, and their uniform size makes planning and patterning surprisingly simple. The finished structure can range from a freestanding wall made of brick alone to a complex affair—a flight of stairs or a backyard barbecue, for instance —consisting of an inexpensive cinder-block core with a handsome brick veneer covering. The freestanding wall, lending charm to the smallest yard or enclosing spaces such as flower beds and play areas, is probably the most common and the most popular. You can build your own—a wall standing on a concrete footing and rising up to 4 feet high—by using the techniques that are shown in the illustrations on the following pages.

The planning of a brick-construction job begins when you choose a location, well before the first brick is laid. For the brick wall, start by consulting local ordinances, building codes, your neighbors and (if you do not own the property)

your landlord, to be sure there are no legal obstacles to your proposed wall. Next, check your soil for drainage; the best-built wall may buckle or sink if it rests on marshy or spongy ground. Study the exact site of the wall with special care: A hill or slope presents special difficulties; avoid large trees with thick and widespread roots; and make sure that the concrete footing of the wall, which will extend about 2 inches to the front and rear, will not overlap an adjacent property line or sidewalk.

Now plan the wall itself in detail. To avoid the complexities of reinforcement rods you should keep the wall under 4 feet high, but you can make it as long as you like and shape it with square corners. You can choose among a number of pattern bonds—that is, different ways of interlocking the bricks—but you are probably best off with the simplest pattern, called running bond, in which the bricks overlap one another so that vertical joints are staggered from course to course. (The basic wall shown in these pages is laid in running bond.)

When you have decided upon the size and shape of the wall, estimate the amount of materials you will need and bring them to the site all at once. An 8-by-12-inch concrete footing contains

6.66 cubic feet of concrete for each 10 feet of its length. To determine the number of bricks, multiply the length of the wall by its height, double this figure (because the finished wall will be two bricks thick), and multiply by 7.5. Plan on about 1 cubic foot of mortar for each 5 square feet of wall surface.

When you place your order at a building supply dealer, select bricks that are roughly twice as long as they are wide, so that the topmost course, when placed across the parallel rows of bricks below it, will cover them completely. Be sure that the dealer has more of the same bricks on hand—bricks are often delivered broken or chipped, and you may spoil more bricks than you expect when you have to split them so that they will fit the running-bond pattern.

When you have completed the footing, let the concrete dry a day or so before going on to bricklaying. As you work, set separate piles of bricks at convenient points to save time, and keep a bucket of clean water or a hose nearby to clean your trowel and your level. Wet down the completed wall, and keep it moist for several days as the mortar cures. After two weeks or so, clean off all mortar stains and any areas of efflorescence that may have developed (page 37).

Making a Story Pole

To control the heights of courses in any brick structure, use a homemade measuring stick called a story pole. For a brick wall, cut a piece of scrap lumber to the total planned height of the wall. With a laundry marker or a similar indelible marker, draw a line near one end of the pole to indicate the top of the bricks in the first course. This first mark on the story pole should be equal to the combined height of a ½-inch mortar bed plus the exact height of a single brick (usually 2¼ inches). Then mark the brick height and mortar bed of each successive course all the way up the pole. As you build the wall, set the pole against newly laid bricks to make sure that the courses of brick rise evenly at every point.

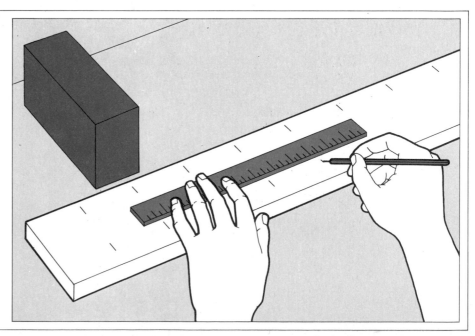

The anatomy of a brick wall. A freestanding brick wall, shown here in cross section, is actually a combination of simple structures. Like all walls, it rests on a footing—in this case, a 6-inch-deep bed of gravel, topped by a poured concrete footing slab 8 inches deep and 12 inches wide.

The top of the footing is about ¾ inch below ground level. Some walls consist of a single row of bricks; for added strength and better proportions, this one has two parallel rows, separated by a narrow air space and bound together at regular intervals by metal strips called wall ties. Both

rows are formed by stretcher courses—bricks placed end to end, with ½-inch-thick vertical and horizontal mortar joints. At the top, the rows are locked together by a rowlock course —bricks set on their long narrow sides and extending from the front of the wall to the back.

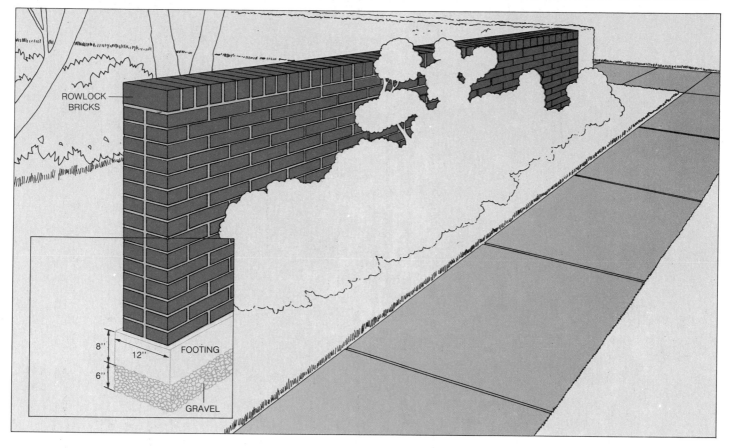

ROWLOCK BRICKS

8"
12"
FOOTING
6"
GRAVEL

The Plan

A dry run for the first courses. From a fixed horizontal reference line, such as the side of your house, driveway or property line, measure out to the baseline you have chosen for the face, or front, of the wall. Drive stakes at the ends of this line and stretch a string between them. Then lay out the first face course of bricks on the ground between the stakes, following the string as a guide and using your forefinger to make ½-inch gaps between bricks. If the end bricks do not quite reach the stakes or fall slightly beyond them, move the stakes to fit the bricks.

Begin the rear, or backup, course of bricks about ½ inch behind the face course starting with a half brick *(page 17)* and continuing with full-length stretchers. When you have placed several bricks, set a rowlock brick across the parallel rows; if it does not fit exactly across the bricks, adjust the width of the space between front and back courses. Lay out the remaining backup bricks, using a second half brick for the far end.

The Footing

Building and marking the footing. Mark the baseline on the ground with sand (*page 43*) and use it as a guide for a trench about 14¾ inches deep. The trench must be long and wide enough to extend 2 inches beyond the wall in every direction: dig it 2 inches beyond the ends of the baseline and about 12 inches wide, overlapping the baseline 2 inches in front and 10 behind. Fill the bottom with a layer of gravel 6 inches deep, then assemble forms, mix concrete and pour the footing to ¾ inch below ground level. In the final smoothing, taper the edges of the concrete downward very slightly—no more than ⅛ inch—to provide a runoff for excess moisture. Finally, mark the baseline again, 2 inches from the front of the footing, with a chalk line (*drawing*).

CHALK LINE

2"

The Lead

1 **Laying the first bricks.** Hose down about 25 bricks or immerse them in water for about 45 minutes, then let the surface moisture evaporate. (Follow this procedure each time you use more bricks.) Mix a cubic foot of mortar (*page 12*). Moisten about three feet of the surface at one end of the foundation with a hose set to a fine spray and let this surface moisture evaporate. Throw a mortar line just behind the chalk line and lay up three bricks on the mortar bed (*pages 13 and 14*). These first bricks will begin to form the lead, or end of the wall.

To make sure the bricks exactly follow the chalk line, align a level from the bricks to the chalk line beyond them. Adjust the bricks, if necessary, to make them perfectly straight and flush to the chalk line. Set a story pole beside the bricks at various points to be sure that the bed joint measures ½ inch; the top of the brick should align with the first mark on the pole. (Repeat these story-pole checks as each course of brick is laid.)

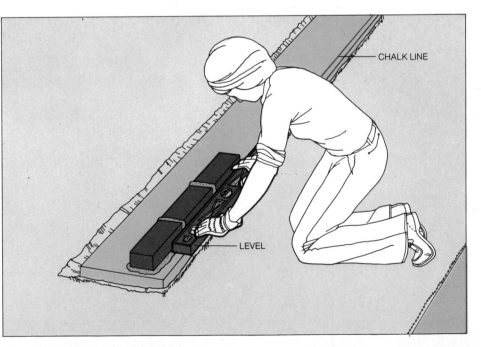

CHALK LINE

LEVEL

2 **Starting the backup course.** Throw a mortar line parallel to the three bricks you have laid for the face course and about ½ inch behind them. Place a half brick at the end of the mortar line, ½ inch behind the first face brick—or at a similar spacing determined in your dry run—then continue the backup course with two stretcher bricks. (As the wall rises, half bricks will alternate between the ends of the face and backup courses.) Use a level to align the backup bricks, and set the level across both courses at several points to be sure that the front and back bricks are level with each other.

Start the second courses of face and backup bricks, beginning with a half brick for the face course and a whole brick for the backup; be careful not to let any mortar fall into the space between the courses as you trim it off. Lay two whole bricks on each course, so that there is a step up from the first to second course; these steps will run to the top of the lead.

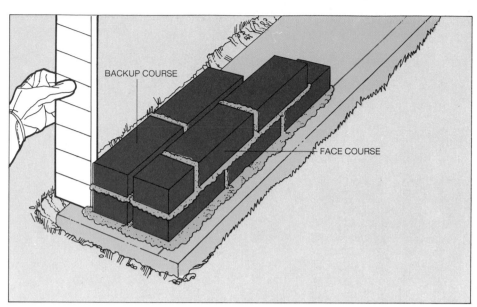

BACKUP COURSE

FACE COURSE

3 **Placing the first ties.** Wall ties (insert) are placed along the wall atop the second course and all other even-numbered courses. Throw a mortar line on the second face course and embed ties in the mortar about 12 inches apart, with the free ends of the ties lying over the backup course (drawing). Then lay two whole stretcher bricks of the third face course over the embedded ties.

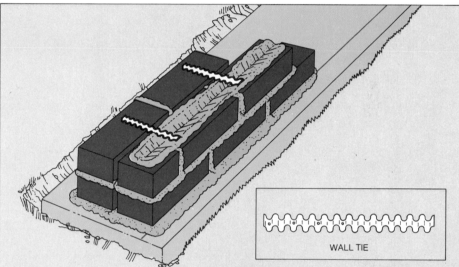

WALL TIE

4 **Mortaring the ties.** When the mortar below the third face course has begun to set, bend the wall ties up from the tops of the backup bricks. Caution: This phase of the job calls for special care to avoid disturbing the bricks already in place. Throw a mortar line on the backup bricks, then bend the ties down into the fresh mortar bed. Lay the backup course—one half brick followed by a whole one.

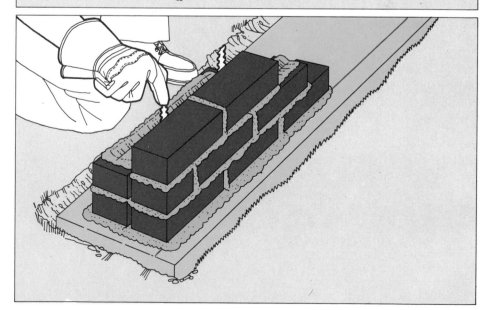

5 **Completing the first lead.** Lay five courses of the face and backup, with wall ties between the fourth and fifth courses; the lead should now be stepped up to a single brick at the end of the face and a half brick at the end of the backup. Use the level to check alignment. If you find a protruding brick, tap the level gently with your trowel handle to push the brick back into line. If a brick recedes, tap it from behind to bring it flush to the level. Do not worry about minute irregularities; correct only those that are obtrusive.

6 **Building the opposite lead.** At the opposite end of the footing, repeat Steps 1 through 5 to form a five-course lead. Checks with the story pole and the level are especially important at this stage of the job. The two leads must match exactly; if they do not, the completed wall will be unstable, with no way to correct the problem short of tearing down the leads and starting afresh.

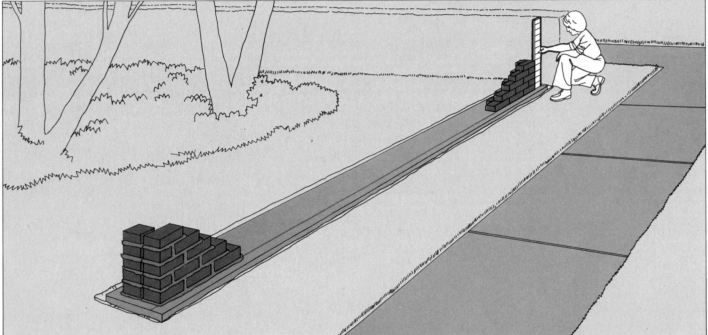

The Corners

1 **Preparing a mason's line.** As a guideline from lead to lead, use a mason's line—a string, slightly longer than the wall, stretched between wood or plastic blocks. Tie one end around a block, then fit it into the lengthwise groove.

Hook the block around the end brick in the first course at either end of the wall, aligning the string precisely to the top edge of the brick. Extend the line to the other end of the wall, fasten the string to the second block, and hook this block around the corresponding brick in the first course; just as before, the string should be flush with the top edge of the brick.

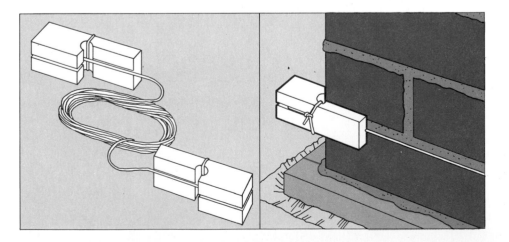

2 **Laying bricks between the leads.** Working from the ends of the wall toward the middle, lay the first face course between the leads, using the mason's line as a guide; the line should be 1/16 inch in front of the bricks and flush with the top edges. At the center of the course, place a closure brick, buttering both ends. Then lay the first backup course, using the line at the back of the wall.

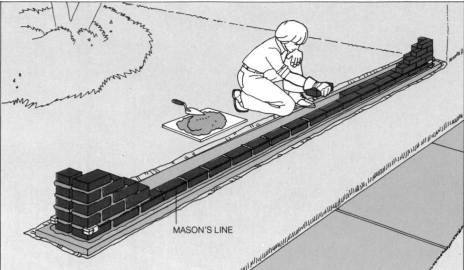

MASON'S LINE

3 **Building up to the top of the leads.** Always working from the ends toward the middle of the wall, lay the next four face and backup courses. Move the mason's line up one course at a time as you proceed, and insert wall ties about every 12 inches atop the second and fourth courses. When you have inserted closure bricks in the last course, the wall will be approximately 14 inches high. If you wish to complete it at this point, lay a cap course of rowlock bricks and fill the joints between end bricks of the two parallel rows with an inch of mortar.

4 **Extending the wall upward.** If you want a higher wall, build new five-course leads at the ends and fill in the courses between the leads, always working from the ends toward the middle. As before, use a story pole as a guideline for the leads and a mason's line for the bricks between them. These five additional courses give a wall 28 inches high; a third set of five courses makes it about 3½ feet high. Two more courses of brick—making 17— raise the wall to maximum practical height.

The Cap

1 Planning the rowlocks. When you have raised the wall to the height you have chosen, cap it with a final course of rowlock bricks, laid across the wall from front to back. Set the course out dry, with gaps for ½-inch mortar joints between the bricks. If the last brick extends beyond the end of the wall, remove it and, allowing for the ½-inch joint, measure the distance from the preceding brick to the edge of the wall. Mark a cutting line around the last brick at this distance, then score and split the brick lengthwise. Place the split brick back on the wall to be sure that the fit is precise, then remove all the bricks.

ROWLOCK BRICKS

SPLIT BRICK HERE

PARTIAL BRICK

2 Laying the rowlocks. Starting at one end of the wall, throw mortar lines on the top face and back-up course to set the first rowlock brick at the end of the wall; then butter one side of the next brick with enough mortar to cover the side completely, and lay this brick with a ½-inch joint (measure this and all succeeding joints to be sure that the course matches the dry run). Continue laying rowlocks along the wall to within three or four bricks of the other end. Because three rowlocks correspond roughly to one stretcher brick, you must make some end-on-end joints, called jack joints, between rowlock and stretcher bricks; in this case they will not affect wall strength.

3 Completing the rowlocks. Insert the partial brick you cut for Step 1 at this point—four or five bricks from the end of the wall—to make it less obvious in the finished course, then lay the remaining bricks. When you have placed the last brick, hold it in place for several minutes until the mortar hardens slightly. Then make a final check of the entire course with a level to be sure that each brick is perfectly aligned in all directions.

A Wall with Corners

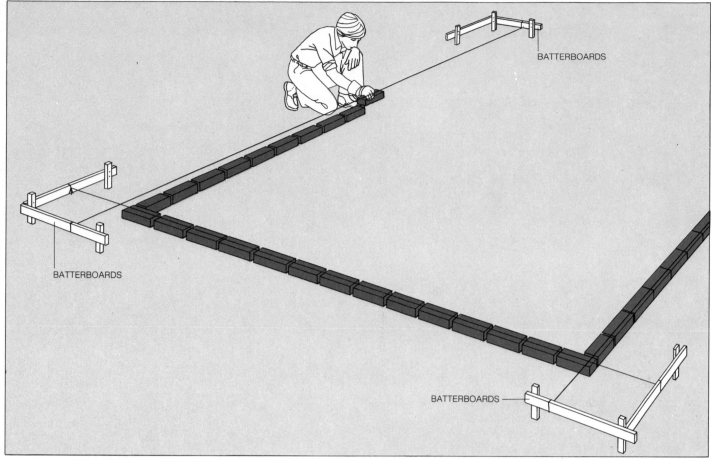

BATTERBOARDS

BATTERBOARDS

BATTERBOARDS

1 **Planning the wall.** Determine the shape of the wall and establish a base line for the most prominent section. Then, following the procedure given for a straight wall, lay out the face course for this section with dry bricks. Form the beginnings of corners with single bricks set at right angles to the main section, using a carpenter's steel square to fix the angles temporarily.

As guidelines for the concrete footings, set up batterboards at the corners and ends of the wall and use the 3-4-5 triangle method to align the batterboard strings flush to the front of the face courses and intersecting at the corners. Use a plumb line to be sure that the intersections of the strings fall directly above the points you have chosen for the corners; readjust the strings if necessary. Then plumb the points for the ends of the wall and mark the strings with chalk or colored fabric to indicate both the lines of the wall and a point 2 inches beyond them for the lines of the footing. Lay the sides of the wall dry (*drawing*), fitting the bricks precisely; then remove the bricks and mark the batterboard strings on the ground by pouring sand over the strings as a guide for the position of the trench for the footing. Mark the position of the strings on the batterboards and take away the strings, leaving the batterboards in place.

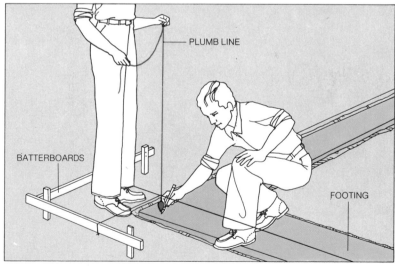

PLUMB LINE

BATTERBOARDS

FOOTING

2 **Marking the footing.** Pour concrete for the footing, with the surface ¾ inch below ground level and the edges slightly tapered. Replace the batterboard strings and drop plumb lines to mark the footings at the corners (*drawing*) and the ends of the wall. On each section of the footing, stretch and snap a chalk line between each of the marks to indicate the position desired for the face course of bricks.

3 **Forming the corner.** Throw two mortar lines at one of the corners, on the foundation, just inside the chalk lines. Lay brick A *(drawing)* on the corner, then butter and lay up brick B, using a steel square to check that the bricks form a right angle. Lay up the four bricks beyond the corner in the order shown—C, D, E and F—to make a six-brick corner lead, and check with a level to be sure the bricks are flush with the chalk line.

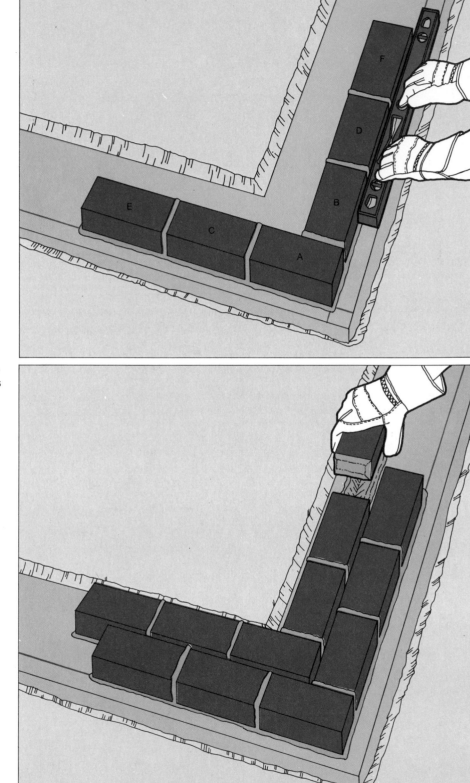

4 **Starting the backup lead.** Throw mortar lines behind the face course, and lay the first three bricks of the backup course *(drawing)* far enough behind the face course to make a perfect fit for the cap course of rowlocks. The backup bricks must form a right angle and overlap the face bricks by ½ brick; use the story pole to measure the height of this and succeeding courses.

5 **Completing the leads.** Working on the front and backup courses simultaneously, build up the corner lead to a height of five courses, with the ends of each course stepped down from the corner. Then build the corresponding corner and end leads or, if you plan to have only one corner, the end leads alone. Stretch a mason's line between the corners—or between a corner and an end —and fill in the bricks between the leads. If you want a wall higher than five courses, add corner and end leads, and lay bricks between them until you reach the height you have chosen.

6 **Laying the rowlocks.** Starting at a front corner of the wall, lay dry rowlock bricks to determine the correct fit of the cap course. If necessary, split a brick at the end. Lay this row of bricks as you would for a straight wall, with the partial brick inserted three or four bricks from the end. Then lay the rowlocks for the other side or sides of the wall, as shown in the drawing at right. Continue until the wall is completed. Cover and cure the completed wall, wait several weeks, then clean off all mortar stains and efflorescence.

Veneering a Concrete Slab with Ceramic Tiles

An elegant way to finish a concrete path or patio is to cover it with a veneer of unglazed ceramic tiles. Strong and durable fired clay, they differ from bathroom wall tiles because they do not have a shiny surface glaze, which might make them slippery underfoot.

To ensure a smooth, professional-looking job, the first step in veneering with tiles is to check and, if necessary, level your concrete slab *(opposite top)*. Since you will lay the tiles in mortar no more than ¼ inch thick, a base with greater variation will make the tiles tilt up and down across the surface.

Forming the thin-set mortar bed, as shown in the cross section below, is a three-step operation that has to be completed without any delays. A paper-thin layer of mortar, or skim coat, is applied with the smooth side of a notched tile trowel; over the skim coat immediately goes a ¼-inch top coat, which finally is raked with the notched side of the trowel to form ridges of uniform thickness. Seat tiles in the mortar with uniform joint spaces between them—usually about ⅜ inch wide. After the mortar cures, put ceramic tile grout into all spaces except those located above expansion joints, which are filled in with polyethylene foam rope and ordinary caulking.

Three types of tile come in unglazed versions suitable for outdoor use: mosaic, paver and quarry tiles. The mosaic tiles, which are laid slightly differently from the remaining two types of tile, are small, usually 1 to 2 inches across and ¼ inch thick, and come mounted in groups on rectangles of paper or mesh. Paver and quarry tiles are 6 or 8 inches across and about ½ inch thick. You can recognize the difference between paver and quarry tiles by the edges around the top faces: on pavers, the edges are slightly curved; on quarries, they form sharp right angles. Because of the angled edges, quarries form a smoother surface than pavers do, and are preferred for floors over which chairs must slide.

How tiles stand up outdoors depends not on size or type but on water absorption. If water penetrates tile and freezes inside, the tile cracks. Only unglazed tiles with a water absorption rating under 5 per cent—your supplier should have manufacturers' specifications that show the figure—can be safely used outdoors in cold climates.

Mosaics are sold by the 1- or 2-foot-square sheet; pavers and quarries by the carton, which may hold enough to cover 10 to 15 square feet. To estimate your needs, multiply the length of the area you want to cover by the width (treating an L-shaped area like two rectangles) and add 5 per cent for waste. In case you overestimate your requirements, most dealers give a refund for unused sheets or unopened cartons.

For laying ceramic tiles outdoors, you need dry-set mortar mix, which comes in 20- to 50-pound bags; 10 pounds are enough for 15 square feet of tile. Ceramic tile grout comes in 5- and 10-pound packages and is available in colors. One pound makes enough for one square foot of paver or quarry tile, two square feet of mosaic tile.

On many jobs, the only tools you need to install tiles are a notched tile trowel to apply mortar, a pointing trowel to pack in grout and a caulking gun to seal the space above expansion joints. If you plan to tile around a flagpole or anything else projecting from the concrete, you can lay mosaic tiles to follow the curves roughly, but you will have to preshape pavers or quarries with a tile nipper or a pair of heavy-duty pliers.

You may need to cut paver or quarry tiles to make them exactly fit the concrete slab and avoid overlapping expansion joints, which must be left free to shift. Make a dry-run layout of two full rows of pavers or quarry tiles at right angles along the edges of the slab. If the tiles fall short of or overlap an edge or expansion joint by less than 1 inch, adjust the spacing. If the shortage or overlap exceeds 1 inch, though, you should rent a heavy-duty quarry tile cutter and cut tiles to fill in the space.

The Anatomy of Tile Paving

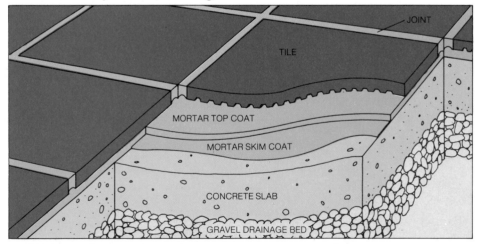

JOINT

TILE

MORTAR TOP COAT

MORTAR SKIM COAT

CONCRETE SLAB

GRAVEL DRAINAGE BED

The structure of tile paving. Tiles are placed in mortar over an absolutely smooth concrete slab. The slab rests on a gravel drainage bed and slopes away from the nearest wall so water will flow off. The mortar bed is made of two layers: a skim-coat layer that bonds mortar to the slab, and a thicker top coat that holds the tiles.

Establishing a Sound Base

Smoothing the surface. Check the flatness of the slab by slowly rolling a long piece of pipe over the surface while, from a low angle, you look for slits of light under the pipe. Use a crayon or chalk to outline areas where light slits indicate irregularities of more than ¼ inch. Then, wearing goggles, flatten the high spots with a rub brick (*drawing*) or a rented electric concrete grinder. Fill in depressions and holes by troweling fast-drying cement-sand-epoxy compound over them; level the patches with a straightedge. Repair cracks or spalling (*pages 20 and 21*). Then hose the concrete clean.

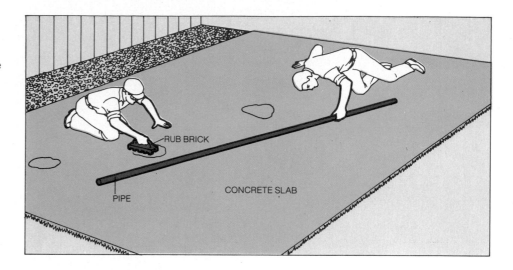

RUB BRICK

PIPE

CONCRETE SLAB

The Mortar Bed

1 Mixing the mortar. Pour 1 cup of water into a dishpan and sprinkle 3 cups of dry-set mortar mix over it—enough to cover 4 to 6 square feet. Stir with a trowel until the mixture is smooth. Then slowly add more mix or more water until the mortar is of a consistency thick enough to form peaks when you pull out the trowel.

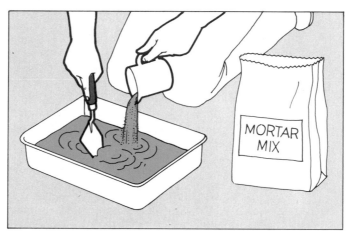

MORTAR MIX

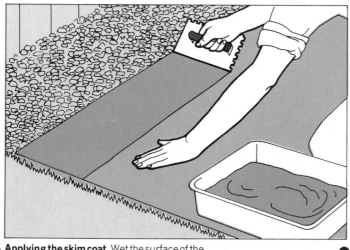

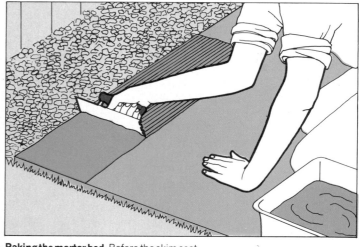

2 Applying the skim coat. Wet the surface of the slab with a broom soaked in water. Then slice into the mixed mortar with the long smooth edge of the trowel and scoop about half a trowelful of mortar onto the underside of the blade. Set the smooth edge of the loaded blade where you want to begin the mortar bed. Holding the trowel at a 45° angle while pressing down firmly, spread mortar paper-thin, coating an area about 2 to 3 feet long and 1 foot wide.

3 Raking the mortar bed. Before the skim coat hardens, scoop more mortar—this time a trowelful—onto the underside of the blade. Again using the smooth edge of the blade, spread a layer of mortar ¼ inch thick over the skim coat. As soon as the skim coat is completely covered, use the long-notched edge of the trowel to rake lengthwise through the mortar, pressing down against the surface of the concrete slab to form a bed of uniformly thick ridges.

Laying Pavers and Quarry Tiles

1 Seating the first tile. Position the first tile, face up, at the corner, aligning the outside edges with the sides of the slab. Spread your hand over the tile's face, and push the tile downward into the mortar bed to seat it. Then tap the tile with the wood handle of a trowel to force mortar up between the ridges on the underside of the tile.

2 Laying adjacent tiles. Leave uniform spaces for grout between the tiles as you proceed to seat and embed them. Lay tiles all the way across the mortar bed and then work in parallel, width-wise rows to cover all the mortar. Pavers made with projecting spacer lugs will automatically line up an equal distance apart when you abut the lugs. To space tiles lacking lugs, use a strip of wood ⅜ inch wide (or of the width determined by your dry-run layout) as a gauge.

3 Truing the tile bed. To check that the tiles lie flat and even, place a long, straight board diagonally across the row after laying the first group of tiles. If necessary, flatten tiles by tapping the board gently with a hammer. Align the outside edges of the tiles while you can still shift them. Repeat Steps 1 through 3 across the edge of the slab, flattening and aligning each group of tiles as you go. Tile up to, but not over, expansion joints. Then work in parallel rows to the edge of the slab that is farthest from any building.

You can stop work at any time, but if you do, be sure to scrape excess mortar from the sides of the last tiles laid. When you have covered the entire concrete base, clean the area as described on page 37. Then let the mortar cure for at least 48 hours before you fill the joints with grout.

A Machine
for Straight Cuts

1 **Scoring the tile.** Set the guide on a quarry-tile cutter to the number of inches you want to remove from the tile. Lift the handle bar and place the tile, face up, against the guide. Lower the bar. Holding the tile steady, press down on the movable handle and push it smoothly away from you to score a straight line across the tile.

2 **Cutting the tile.** Place the tile in the clamp at the end of the cutter with the scored line centered under the clamp. Lower the handle bar, then tap it gently with the edge of your open hand. A clean, straight split will result.

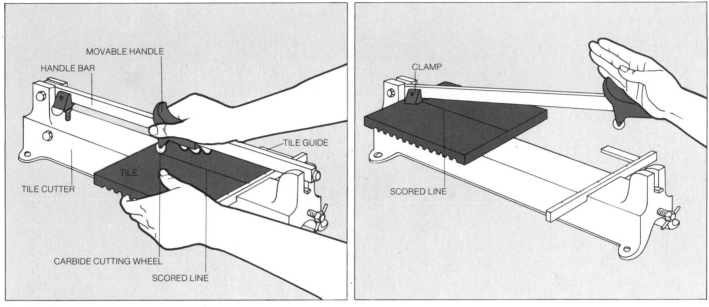

A Nipper
for Curved Cuts

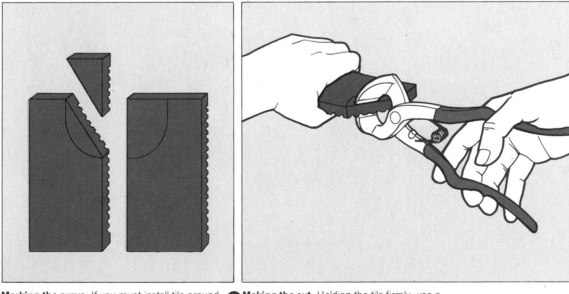

1 **Marking the curve.** If you must install tile around a pipe or other round projection, make a cardboard template of the required cut and trace it onto the face of the tile. Use a quarry-tile cutter to remove most of the unwanted segment. If the curved area falls within a tile as shown above, cut the tile in half, shape each and butt the halves.

2 **Making the cut.** Holding the tile firmly, use a tile nipper or heavy-duty pliers to break off small bits of the excess at a time. Do not expect curved cuts to be as perfect as straight ones. When you finish tiling, you will fill the space along the curve with polyethylene rope and caulk so that irregularities will not be noticeable.

Laying Mosaic Tiles

Installing sheets of tiles. Mosaics come already assembled and evenly spaced on sheets of paper or fabric mesh so you can lay them by the sheet rather than individually. Start by spreading a large enough mortar bed at the corner of the concrete slab to hold one or two full sheets of tiles. Without removing the paper or fabric, seat the tiles at the corner—tile side up if the mosaics are back-mounted (*upper drawing*) and paper side up if they are face-mounted (*lower drawing*). Following the instructions for laying paver and quarry tiles, embed the mosaics and true them. The space between adjacent sheets should be as wide as the space between tiles on the sheet. Continue across the width of the slab, then in parallel rows. At expansion joints or the ends of rows, use a utility knife to cut sheets to fit. If you are using face-mounted tiles, wet the mounting paper with a sponge and lift it off before the mortar begins to set. With either kind of tile, let the mortar cure for 48 hours before grouting.

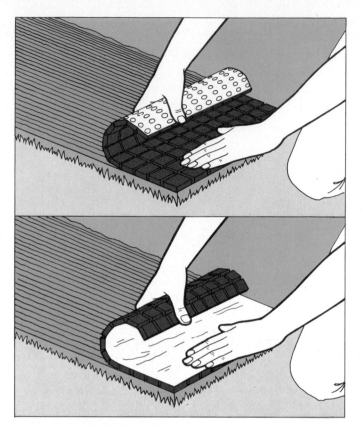

Three Ways to Fill Joints

ROLLED NEWSPAPER

1 **Mixing the grout.** Pour 1 cup of water into a bucket and add 1 quart of ceramic tile grout mix. Stir until the grout is smooth and has the consistency of wet sand. If necessary, add water. Some grout must stand for 20 minutes and then be restirred before use; check the label.

2 **Troweling grout into joints.** Stuff rolled newspaper into the spaces above expansion joints or around curved-cut tiles, to keep these spaces clean. Start work on the slab nearest the building. Slice a trowel full of grout onto the blade of a pointing trowel and set the tip of the blade midway along one joint. Tip the trowel so the grout slides into the joint and forms a mound. Make similar mounds, 3 or 4 inches apart, in as many joints as you can reach.

3 **Leveling the grout joints.** Spread the grout into the joints with a window-washing squeegee or a stiff cardboard. Press hard enough to pack the grout and to force out air bubbles. As you work, push excess grout toward nearby joints.

4 **Cleaning the joints and tiles.** Sprinkle a generous amount of dry grout mix over the grouted joints and rub with dry burlap, using a circular motion. This action compacts the grout and prevents unsightly cavities that might appear as the grout dries and shrinks. Use the burlap dabbed with grout mix to scour the face of the tile. Then sweep the grout dust from the surface.

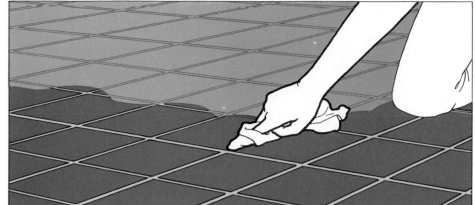

5 **Edging with grout.** To cover the exposed edges of the tiles and slab, use a pointing trowel to place a trowelful of grout on the exposed edge. Then, holding the blade against the top edge of the tiles at a 45° angle, run the tip of the trowel along the ground to spread the grout evenly. Cover the tilework with burlap and sprinkle it for at least three days to let the grout cure.

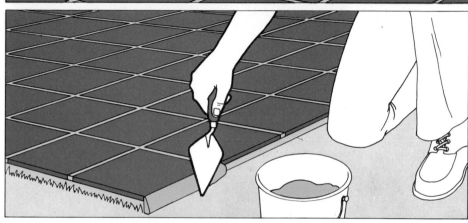

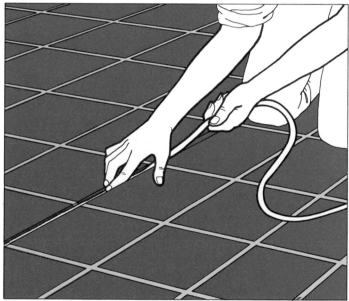

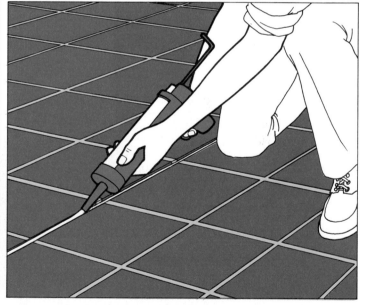

6 **Filling expansion joints.** After the grout is cured, remove the rolled newspapers from the expansion joints and from around curved-cut tiles. Then press into them polyethylene foam rope that is the same thickness as the joints.

7 **Sealing expansion joints.** Use a caulking gun to fill the spaces above the ropes with a self-leveling silicone or polysulfide caulk. Wipe any caulk from the tiles immediately, using the solvent recommended by the manufacturer. Let the caulk dry until it is not sticky before walking on the tiles.

The Roughhewn Appeal of Natural Stone

No masonry material is more handsome or durable than the oldest: rock. It is heavy to work with, but its very weight makes a structure that is solid in look and in fact. A flagstone walk laid dry in sand —without mortar—resists being tipped from frost. And the simple dry walls that crisscross New England, standing after centuries of northern winters, are still a good model for low, home-built walls.

Even stone can shift if not mortared—a dry wall may require a bit of fixing each spring—and maintenance-free stone paving and walls should be laid with grouted joints in mortar beds over concrete bases. The mortar must be thick to prevent it from seeping out of the joints. To strengthen the bond between paving stone and mortar, cement butter—a barely wet mixture of cement and water —must be placed under each stone.

With or without mortaring, the secret of handsome stonework is locking the irregular surfaces of the stones together snugly. How easy this part of the job turns out to be will depend on the form of stone you choose. Natural or quarried rubble produces the most rustic-looking walls; irregular or mosaic flagging gives the most roughhewn paving. Any of these forms requires trial and error to arrange. The square-cut stones *(bottom right)* that create gridded wall patterns and the rectangular pattern flagging that forms geometric paving need less experimentation as you go along.

All three forms—rubble, square-cut and flagging—are produced from a variety of stone types, and which type you use depends largely on what is available from nearby quarries. Granites, which are usually grayish in color, are the hardest and most durable, but also the most difficult to cut and highest in price. Limestones vary in color and in degree of hardness, from very compact to granular. Slates, purple, gray and green, are hard and nonporous; because they are naturally stratified into layers, they are commonly split into flagging. Bluestones and sandstones are available in many colors, from cream and pink to red and dark blue; most are easy to cut and trim.

Since no two stones in nature are ever exactly alike, no two stones you buy will be identical in size, shape or color. You can estimate quantity roughly, however, on the basis of the form of stone you want. Flagging is sold by the square foot; to determine how much you need, measure the area you plan to pave and allow about 10 per cent extra for waste. Rubble and square-cut stones are sold by the cubic yard. To determine how much you need, multiply the number of feet in the length of the wall you want by the width and then the height; divide the result by 27 to determine the total cubic yards, then add 10 per cent for waste.

Because of the irregularity of stone sizes and surfaces, accurately estimating the amount of mortar, grout and cement butter you will need to use is difficult. As a starting point, you can figure about 150 pounds of portland cement and 500 pounds of mortar sand for each 50 square feet of paving or 1 cubic yard of wall.

Before delivery day, clear a large space close to the work site and spread out tarpaulins or plastic sheeting to protect the lawn. For moving stones around, rent a dolly—it reduces the amount of lifting you must do. Standard masonry tools are used, and standard precautions are necessary: lift with your legs, not your back; wear heavy gloves; and put on goggles whenever you trim or split stone.

The Stones a Mason Works With

QUARRIED RUBBLE

FIELDSTONE

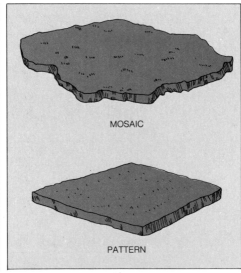

MOSAIC

PATTERN

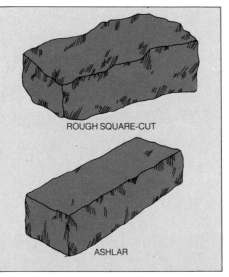

ROUGH SQUARE-CUT

ASHLAR

Rubble for a rustic look. Rubble is uncut stone, quarried as well as natural field and river stones. Most dealers sell only quarried rubble, in pieces 6 to 18 inches in diameter. The rough surface of quarried rubble holds mortar better than the worn surfaces of field or river stones.

Flagstones for paving. Flagging is made by splitting stone in thin slabs. It may be laid in irregular mosaics or cut into patterns. Flags range in size from ½ to 4 square feet in area and from ½ to 2½ inches in thickness. For a sand bed, flagging should be 1½ to 2 inches thick; for a mortar bed, the flags should be ½ to 1 inch thick.

Square-cut for an easy fit. Square-cut stone is roughly trimmed. In more carefully and expensively shaped versions, it is called ashlar. Widths from 3 to 5 inches, heights from 2 to 8 inches, lengths from 1 to 4 feet are generally available.

Laying Flagstones on a Sand Base

1 Fitting stones. Stake the area you plan to cover with flagging, excavate it to a depth of 3 inches and lay down a 2-inch layer of sand, following the techniques on pages 80-82. Starting at one corner, arrange three or four 1½- to 2-inch-thick flags on the sand at a time. Line up the straight edges with the outside of the excavation and fit irregular edges together so that joint spaces will be ½ inch wide. Use a pencil to mark segments of stones to be trimmed for a snugger fit.

2 Trimming small segments. Place the stone on a bare section of the sand bed. Then, wearing goggles, knock off small unwanted pieces by hitting them outside the pencil marks with a bricklayer's or stonemason's hammer. Save the chips to use as fillers between large stones. If a segment is hard to remove, undercut it first by chipping off bits from the bottom edge.

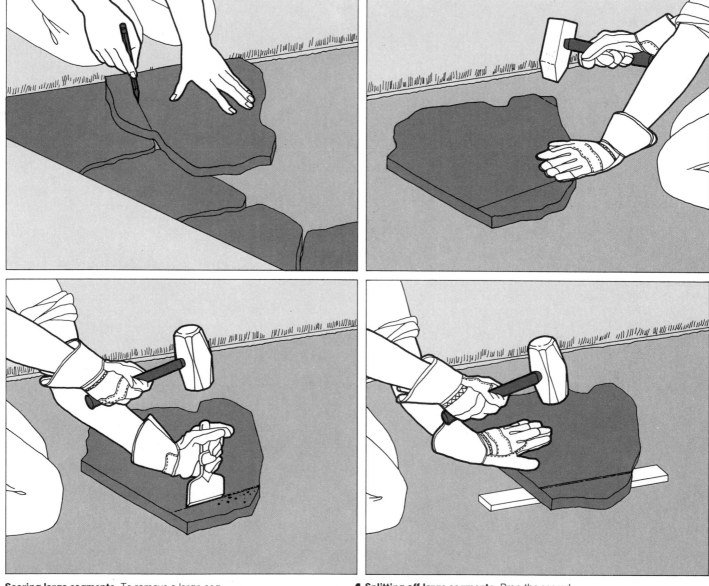

3 Scoring large segments. To remove a large segment of flagging, place the stone on sand and score along the drawn line with a brick set or stonemason's chisel and a heavy hammer.

4 Splitting off large segments. Prop the scored flagging on a board with the unwanted segment tilted upward beyond the edge. Tap the segment repeatedly with a hammer until it falls off. If the stone does not split readily with this treatment, score it along the sides and the back, then prop it up and tap it again.

5 **Embedding stones.** Working from one corner of the paved area, tap each flag down into the sand bed with a rubber mallet. The top of each stone should lie about ½ inch aboveground.

6 **Truing the surface.** After embedding a row or two of flagging, set a mason's level on top of the stones to see if their surfaces are even. Put additional sand under low stones, and scoop out sand from beneath high ones.

7 **Filling the joints.** Shovel additional sand over the flagging and sweep it across the stones until the joints are filled to the brim. Water the surface and let it dry. Repeat the process until the joints are completely filled and compacted. To discourage weeds, sprinkle herbicide into the joints.

Mortaring Flagstones onto a Concrete Base

1 Making a dry run. If you do not have an existing base, lay concrete following the instructions for making a lightweight slab. Let it cure for one week. Arrange ½- to 1-inch-thick dry stones on the base with no more than ¾-inch spaces between them. Trim the stones as you go. Prepare stiff mortar by mixing one part portland cement to four parts mortar sand and a minimum of water —only enough to make the mortar hold the shape of a ball when you grasp it. In a separate container make cement butter: portland cement containing enough water to give it the consistency of soft butter. Then remove three or four stones and wet the concrete with a damp brush.

2 Embedding the stones. Trowel a 1-inch-thick layer of mortar over the wetted area. Put back the stones and tap them down about ½ inch into the mortar with a rubber mallet. When you have embedded a dozen stones or so, level them.

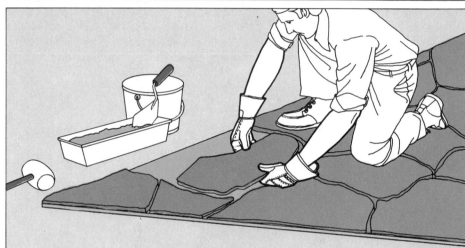

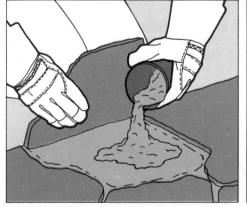

3 Applying cement butter. Immediately after leveling the stones, pick up one flag at a time from the mortar bed and use a paper cup to dribble 2 or 3 ounces of cement butter into the depression it leaves. Replace the stone and tap it back into position with the mallet. When you have buttered the stones, recheck the level.

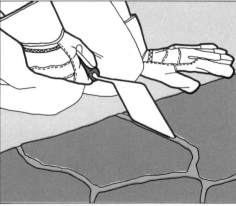

4 Raking the joints. Using the pointed end of a trowel, pack the mortar under the edges of the stones. Then scrape out the excess so the bottoms of the joints are about at the level of the bottoms of the stones. Sponge mortar off the flagging. Let the mortar cure for 24 hours.

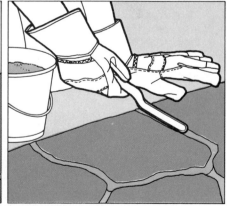

5 Grouting the joints. Prepare a thick grout, using one part portland cement to one part mortar sand with just enough water to make the mixture spreadable. Use the tip of a concave jointer to push prepared grout into the spaces between the flags and compact it. Sponge off the excess, and let the grout dry for 24 hours before walking on the stone paving.

The Tricks of Making a New England Dry Wall

You can build mortarless stone walls like those that have marked off New England fields since colonial days if you follow the simple procedures of the early settlers. Dry walls need neither mortar nor concrete footings since the weight of the stones and their interlocking placement hold them together. If you have never built one before, it is best to limit yourself to a wall 2 to 3 feet thick and no more than 3 feet high.

You can use stones picked up from fields, but they may require considerable cutting to make shapes that join securely. Most builders buy quarried rubble, which has more or less rectangular faces. Easily workable types—bluestone, sandstone or limestones—are best.

A dry wall is built directly on the ground or, for better drainage, on a bed of sand. It has overlapping joints like a running-bond brick wall. But it also has several unique features. One is the bonding stone, the first stone of the first course. Ideally it should be as long as the wall is thick, since it is placed crosswise to anchor the wall, tying the course on the front to the course on the back.

The front and back courses of stones do not make a solid wall. Small stones are used to balance the large ones, the space between the courses is filled with more small pieces, and gaps in front and back are filled by hammering "chink" stones into them.

The faces and ends of a dry wall are not vertical but slope slightly inward from a broad base, because each successive course is inset slightly from the one beneath. This taper can be judged by eye, or with a homemade device called a slope gauge (below, left).

Although the bulk of a dry wall contains no mortar, many builders lay mortar on a top of broad flat stones. This mortared cap seals out water that may freeze and dislodge stones.

While building, observe a few rules that have been followed by generations of New Englanders: tilt stones downward toward the center so that the gravitational pull against the stones will compact the wall and help to keep it intact. When turning corners, avoid mitered joints —those with angular stones butted together at the ends. Never stack joints one above the other; always overlap. And do not be too meticulous; a rough wall generally looks better—and is sturdier—than a fussily even one.

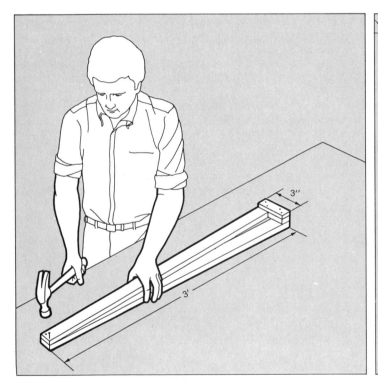

The slope gauge. Each end and face of a dry stone wall must taper inward. To gauge this slope, nail together two 1-by-2s at one end. Complete a right-angle triangle by nailing a short piece across the two 1-by-2s and supporting one end with a small block to keep the long pieces flush. This short piece should measure 1 inch for each foot of wall height.

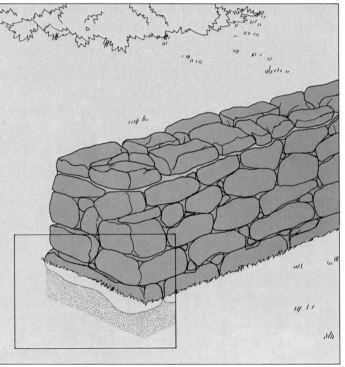

The anatomy of a dry wall. A dry wall sits on a 5-inch layer of sand in a 6-inch trench. At the base is a bonding stone that goes from front to back and is overlapped by other stones. Unlike bricks or square-cut stones, which are laid flat, rubble pieces tilt toward the center of the wall. As each new course is added, it is inset so the entire wall tapers slightly. The top layer of stones is mortared to seal out water.

A Wall for All Seasons

1 **Setting the bonding stone.** Dig a 6-inch trench the length and width of the wall, then fill with about 5 inches of sand. Pick an even-faced stone that is as long as the wall is thick and place it at the end. This is your bonding stone—it helps hold the wall together. If you cannot find a long enough stone, make the bond with two stones.

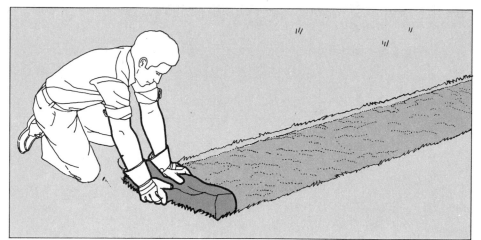

2 **The first course.** Lay stones along one side of the wall, then the other, alternating large and small stones, thick and thin ones, and placing long ones lengthwise, not across the wall thickness. Lay each stone flat, never on end or side, but set so that any slope of the upper surface angles downward into the center of the wall. Use the biggest stones for this first course, saving smaller ones for later courses —and the flattest ones for the top.

3 **Filling in the center.** After you have laid 8 to 10 feet on both sides of the first course, fill in the center with small stones, building it up until the course is reasonably level.

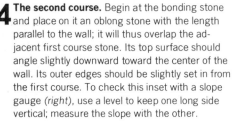

4 **The second course.** Begin at the bonding stone and place on it an oblong stone with the length parallel to the wall; it will thus overlap the adjacent first course stone. Its top surface should angle slightly downward toward the center of the wall. Its outer edges should be slightly set in from the first course. To check this inset with a slope gauge (right), use a level to keep one long side vertical; measure the slope with the other.

Continue laying second course stones along one side, then the other. Choose stones of a length that will overlap joints in the first course, of a shape to conform approximately to the surface underneath, and of a thickness to keep the top of the course approximately level. Check the slope each time. Then fill in the center.

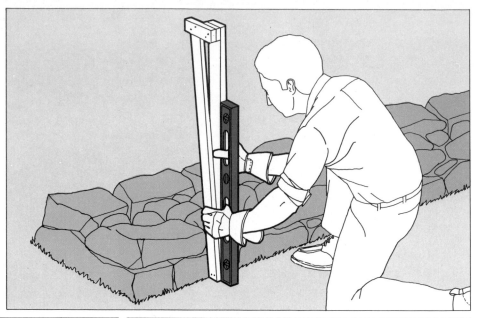

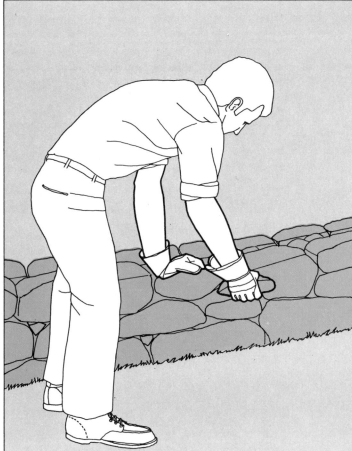

5 **Shimming.** The stones will not always seat firmly on those underneath. Teeter the ends of each as you set it. Insert stone chips or small rocks as shims under the front or back edges until the stone rests securely and its weight tilts inward.

6 **Chinking.** After two or three courses have been laid, fill gaps between stones by driving narrow stones in with your hammer. This chinking process locks the wall tight and helps keep weight pressing inward from both sides. As you lay up the later courses, use the slope gauge to check the inset; judge the horizontal level by eye.

7 **Mortaring the final course.** If you want to apply a mortared cap, cover the next-to-last course with a 1-inch layer of mortar, using a mix of 3 parts sand to 1 part cement. Then set in the stones you have been saving for the top. Fill in the gaps between these stones with mortar, building up the center of the joint to prevent pockets where water might collect. Trim excess mortar from the sides of the wall with the trowel.

Interlocking Stones for a Corner

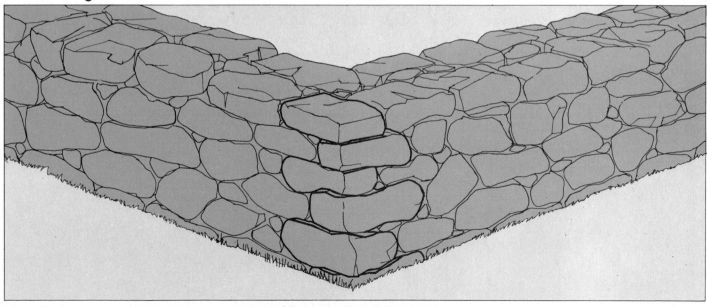

Intersecting the corner stones. Lay stones up to the corner as you would if you were building a straight wall. But at the corner, for the last stone in the first course of the inner face, use a large stone that will overlap the first inner course of the turn. Lay the outer first course the same way. In the second course, set the last stone short of the corner so that its end will meet the side of the first stone around the turn. Overlap succeeding courses in the same way.

The Wet Wall

If you want a more regular and more maintenance-free wall than New Englanders build, use stones that are at least roughly rectangular, not rubble, and mortar them together over a 6-inch-thick concrete base. This "wet" construction is often used for retaining walls, which require weep holes and gravel fill to drain off water.

For a low wet wall—no more than 3 feet high—observe a few variations on the building techniques used for rubble walls. A footing is needed. An entire first course is laid down—this time with mortar surrounding each stone—and later courses are added with the stones overlapping the joints below like a running-bond brick wall. Keep all stones level. And do not build as massively as with rubble; because mortar rather than the weight of the stones holds a wet wall together, it need be no more than 18 inches wide.

To preserve the somewhat irregular contours of the stones, the final raking, or trimming, of the mortar joints is generally deeper than with bricks. Make sure your mortar mix is correct—3 parts sand to 1 part cement is best—and that mortar beds are thick and fluffy. As soon as the mortar is troweled onto the bed (or footing), it is ready for the stone: no furrowing is required.

A Wall Bound by Concrete

1 Laying the mortar bed. Dig a trench for a concrete footing that will extend about 5 inches beyond the base of the finished wall on all of its sides. For a terrace wall, dig out about 2 feet of earth behind the wall to allow for the gravel fill. Pour the footing and let it cure for at least 48 hours. Then put down a 1-inch-thick layer of mortar on the slab, spreading only enough at a time to keep ahead of your stone setting. After throwing the bed, pat, but be careful not to compact, the mortar; it should have a fluffy texture.

Anatomy of a terrace wall. A low wall 16 inches wide sits on a 6-inch gravel bed and a 6-inch-thick concrete footing. To lead away water from the soil behind it, a retaining wall must have weep holes at the bottom, and gravel fill behind to enable water to seep down to the weep holes.

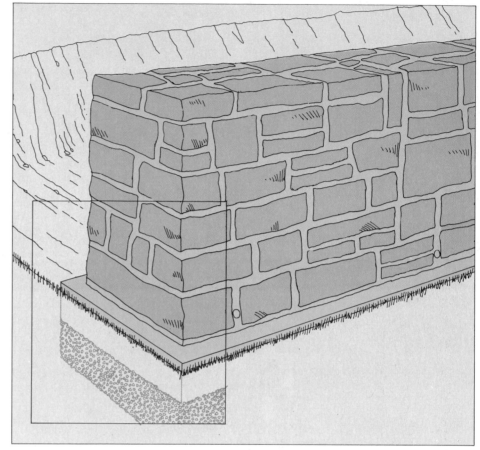

2 **Setting the bonding stones.** Set into the mortar a bonding stone that stretches from front to back, centering it on the slab. Add stones along one side of the first course of the wall, leaving a ¾-inch gap between stones for the mortar. Tap the stones down into the mortar, striking them with the heel of the trowel. If you find, after placing a stone in position, that you have made a mistake and must reset it, first wash off old mortar.

3 **Filling the center.** After you have laid stones along both front and back sides of the first course, fill in the center with odd stones—size and shape are not crucial—and cover each with mortar, adding stones until the course is level. Throw mortar into the spaces between the outside stones. Then lay the sides—but not the center—of the second course similarly, using a slope gauge to determine the correct inset.

4 **Making weep holes.** After laying, but not filling in, the second course, drive a broomstick handle into the wet mortar between stones of the first course. Push the stick through the wall. If it meets a center rock, reach in and remove that piece so the stick can continue, or pull out the stick and try another spot in the wall. When the stick has reached the other side, wiggle it a bit to make sure it has made a clean hole and pull it out. Repeat every 3 or 4 feet. Then fill in the second course and lay up additional courses.

5 **Raking the joints.** Complete the wall by laying the flattest stones on top and mortaring them in as with the dry wall. Rake the joints between stones with a piece of wood, compacting and removing enough mortar so that it is not obtrusive. After the mortar has set for a day or so, fill in behind the wall with gravel to within 6 inches of the top. Then cover the gravel with topsoil.

Hollow Blocks for Economical Construction

Concrete block and cinder block give the mason two major advantages: economy and speed. A block wall 8 inches thick costs about two thirds as much per running foot as an equivalent brick wall and takes about half as long to build. The gain over stone is many times greater.

Both concrete and cinder blocks are cast from concrete—portland cement and fine aggregate—but the cinder type has coarser aggregate (actually including cinders in some areas, but not all), is more porous and weighs less, which makes it easier to handle. A typical block weighs about 30 pounds and measures 8 inches by 8 inches by 16 inches, but there is a wide variety of shapes and sizes for special purposes (below).

Blocks are laid like bricks. Both are anchored to a concrete slab or footing with a furrowed mortar bed, but mortaring the subsequent courses of block differs slightly from brick technique, and some types require steel reinforcement. Also, mortar for block should be stiffer than for bricks; simply cut the water in the mortar recipe for brick until mortar needs to be shaken from the trowel, rather than just slipping off it, as with the brick mixture.

The methods shown here apply to walls of any height—a freestanding garden wall can rise as high as 12 feet when made of 8-inch blocks, 9 feet when made of 6-inch blocks—but skill and experience are necessary if tall walls are to be safe, and beginners are advised to make their first efforts low. Walls under stress must be lower yet.

Because block is economical, it is often used to make a core for what looks like a structure of stone or brick. A brick veneer can be mortared directly over a completed block structure. Or you can build up the block and the casing at the same time, anchoring the facing material to the blocks with wall ties. This method is particularly well suited to building a barbecue, for it leaves a cavity between the block and the sheath so that they can expand independently when the barbecue is used. Their expansion rates differ, and a bond between the two materials might cause breakage.

A Block for Every Spot

Choosing the right block. The stretcher block is the most common. It has mortar-joint projections at both ends and two or three hollow cores separated by partitions called webs. A variation of the stretcher, called a corner block, is flat on one end for use where the end will be exposed. Half blocks laid at the beginning of a wall set up a running bond pattern of mortar joints; you can purchase half blocks ready-made or cut them yourself from corner units. Partition and half-height blocks are handy shapes for tailoring a block core to fill a brick casing and can be combined to make weep holes to drain water from behind a retaining wall.

Where cores of blocks would be visible, solid-top or coreless units are often used in place of hollow ones or are laid on top of them. However, screen blocks, used for light and airy walls and partitions, have cores that are decorative and are meant to be seen.

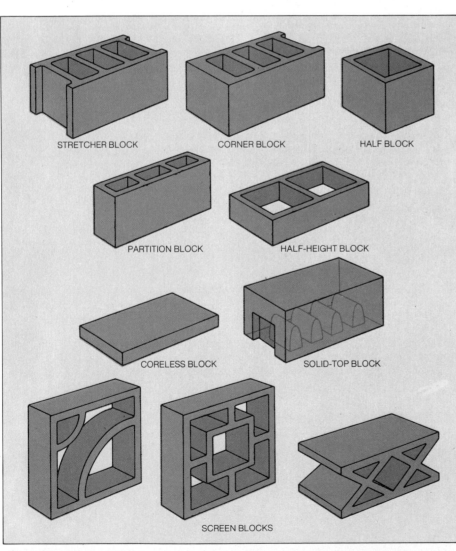

STRETCHER BLOCK   CORNER BLOCK   HALF BLOCK

PARTITION BLOCK   HALF-HEIGHT BLOCK

CORELESS BLOCK   SOLID-TOP BLOCK

SCREEN BLOCKS

How to Lay Blocks

Mortaring blocks to each other. Stretch a mason's line between the leads of the wall—constructed like brick leads. Trowel a two-block-long mortar bed onto the tops of blocks so that 1½ inch ridges of mortar cover the core webs as well as the edges of the blocks. Do not furrow mortar for ordinary blocks, although you may have to for screen blocks. Stand two blocks on the ground and butter the mortar joint projections at one end of each. Lift each buttered block by its webs, and in one motion push it into the mortar bed and against the adjacent block to make 1½-inch joints. Trowel away oozed mortar.

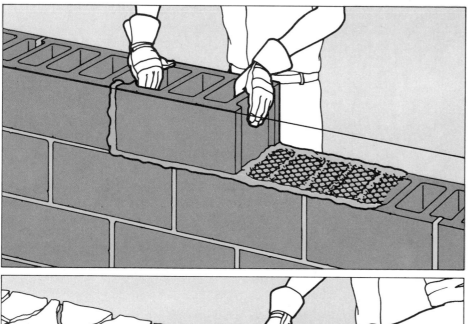

Rainproofing a Block Wall

Filling the cores. To finish the top course of a block wall, use coreless blocks, a coping of stone or brick, or fill the cores in the top course with mortar. A floor of metal mesh under the top-course block keeps the fill from dropping to the bottom of the wall. The mesh is metal lath or hardware cloth cut with metal shears into strips two blocks long and ½ inch wider than the cores. Throw a mortar bed for the top course and push the edges of the mesh strips into the mortar; then lay blocks on top. Finally, trowel mortar into the cores so the fill is even with the tops of the webs.

Attaching a coping. A coping of solid concrete blocks can be laid directly onto the top course of the wall. For a stone coping, however, you must first fill the cores of the top course with mortar and then lay ½-inch mortar bed. Set the coping on the mortar and fill in the joints with grout, as illustrated for a stone walk. If you want the coping to slope, simply spread the mortar bed thicker along one side of the wall than the other.

A Decorative Block Screen

1 Laying the blocks. Screen blocks are laid in a block bond pattern, that is, one in which the vertical mortar joints line up instead of being staggered as they are in the running bond pattern. Start a wall by building a two-block lead at each end, laying one block atop the other. Then, using a mason's line as a guide, fill in the first two courses of the wall by buttering one side of each block and setting it on a mortar bed furrowed with the point of a trowel.

2 Reinforcing. In a decorative block wall, strengthen the mortar bed between every other course with a welded steel lattice called joint reinforcement. It comes in 4-foot and 8-foot lengths and a variety of widths—order it ¼ inch narrower than the block it will rest on. Trim the reinforcement with a hacksaw so that it is about 2 inches shorter than the wall. Throw the mortar bed for a lead and furrow it, then push the joint reinforcement into the mortar before placing the block. When filling in between leads, lift reinforcement into the middle of the mortar after furrowing it.

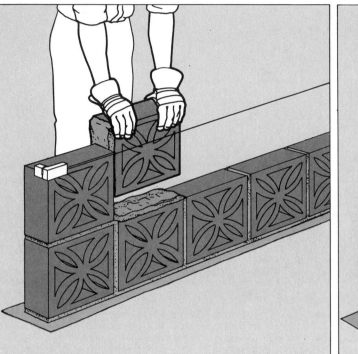

A Cinder Block Retaining Wall

Anatomy of an earth dam. To hold back the weight of soil, a block wall must be strengthened with buttresses called pilasters and provided with superior drainage. In the cutaway drawing at right, cinder-block pilasters are laid on concrete tongues of the footing, set at the ends and every 10 feet of length. The pilasters, one block lower than the wall, are hidden from view by the back fill. For drainage, gravel banked against the wall allows water to seep quickly downward to weep holes. A top course of coreless blocks or stone coping will help to waterproof the wall.

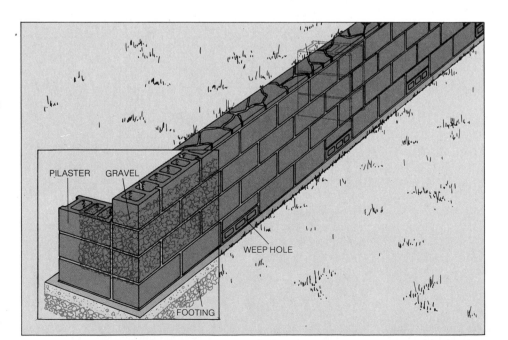

PILASTER GRAVEL

WEEP HOLE

FOOTING

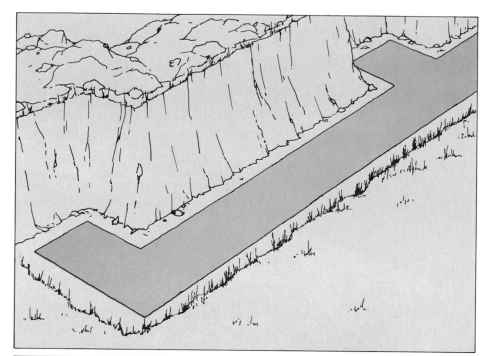

1 Excavating the site. To build a retaining wall in a slope, dig a level space about 2 feet wide, making 3½-foot niches for pilasters. Pile the dirt above the excavation, so it will be easy to fill in afterward. Dig a 12-inch-deep trench for a footing no closer than 6 inches to the edge of the excavation, making the trench about 8 inches longer than the wall and about 16 inches wide with 16-inch-square pilaster tongues every 10 feet and at the ends. Construct a 6-inch concrete footing on a 6-inch gravel base, and reinforce it with 6 inch by 6 inch, 10-gauge steel mesh.

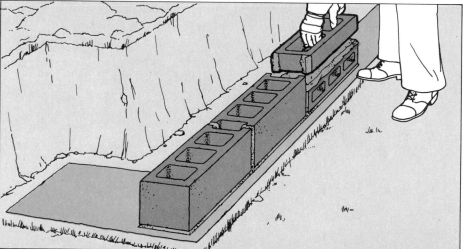

2 Building a lead. Begin with a corner block. Lay a stretcher next to it, then a partition block laid on its side so that the cores run through the wall for weep holes. Lay a half-height block on top of the partition block to maintain the height of the first course, buttering the half-height block end only where it will abut the mortar joint projections of the stretcher block.

3 Tying in pilasters. As you complete each course of the lead, lay a stretcher block at right angles to each end of the wall to make pilasters there. After the second course of the lead, reinforce the joint between pilaster and wall with metal lath or hardware cloth 7 inches wide and about 16 inches long. Throw two mortar beds—one for the third course of the lead and the other for the third pilaster course. Then push the metal mesh into the mortar (*left*), ½ inch from the front edge of the wall.

Fill in between leads, building and tying in pilasters in the same way on the footing tongues provided. Install weep holes every third block in the first course, but where a pilaster coincides with weep holes, simply build the drains one block farther along the wall.

A Block Staircase

Masonry steps made of a block shell filled with sand and rubble are easier, quicker and less costly to build than poured concrete steps. The first stage in constructing such steps is to pour a 4-inch-thick reinforced concrete slab for them to rest on.

Make the slab 4 inches wider and 2 inches longer than the steps and landing. A 40-inch by 40-inch landing, for example, requires a slab 44 inches wide. The slab length would be 42 inches plus 10 inches for each step since a 10-inch tread is a perfect fit for the 8-inch riser given by ordinary cement blocks. Slope the slab ¼ inch for each foot of length so that when finished the steps tilt slightly down from the house. They will then drain well and will also be easier to climb. Place an expansion joint between the slab and the house foundation.

To improve the appearance of block and poured concrete steps, many home-owners paint or veneer them. Masonry paint makes pores and mortar joints less prominent, and stucco—a cement plaster —hides them completely. A veneer of brick, tile or stone is most decorative.

You can attach a masonry casing to existing steps or to ones you have built yourself. If the steps are already there, you will probably have to fit the material to them by trimming it and by widening or narrowing mortar joints. A casing also makes the first step higher than the others, a problem most easily solved by paving the walk leading to the steps.

If you build steps like those shown here and on pages 68-69, you can solve the fitting problems in the planning stages. Start with a slab wide and long enough to support the veneer. Pour the slab below grade to make the first step equal in height to the rest. You can also retain ½-inch mortar joints and eliminate most trimming if you tailor the tread depth of the steps to the casing material.

Building the Steps

1 Laying the side walls. After the concrete slab for the steps has cured overnight, use a chalk-line and steel square to mark the outline of the steps on the slab. Begin at the front edge with a solid-top block so that no cores will be visible in the finished steps. Use stretcher blocks to extend the first course, leveling each block as you lay it. End the course with a corner block, trimmed to fit if necessary. Start the second course with another solid-top block, leaving a 10-inch tread for the first step. Complete one side wall in this way, then the opposite one.

2 Filling in the treads. Stretch a mason's line between the side walls and throw a one-block, furrowed mortar bed onto the slab. Butter the edges of a corner block (not the cross webs in this case), turn the block on its side to hide the cores and lay it against the side wall. Complete the first tread with corner blocks and after it has set for an hour, fill in the space behind with scrap masonry and sand. Tamp the fill even with the top of the first tread.

For the other treads, reposition the mason's line, throw a mortar bed 6 inches wide along the back of the preceding tread, and repeat the first-tread procedure. Fill in the landing at the top with stretchers laid on their sides.

HALF BRICK

Sheathing Steps in Veneer

1 Encasing side walls. Butter the bricks and mortar them to the sides of the block core, following the treads and risers of the steps as closely as possible. The casing should extend far enough in front of each riser and above each tread and the landing so that the veneer material will align with the side wall casing. To achieve this fit with existing steps, adjust the thickness of the mortar joints or cut the casing material.

If you build steps that you plan to sheath with brick, like those illustrated, minimize cutting by making the tread depth 1½ brick lengths plus the dimensions of two mortar joints. Start the first course of the side casing with a half brick, positioned so that the next brick in that course will be even with the front of the block case. Lay four courses with the running bond pattern, positioning each brick with a story pole and level. Start the next three courses 1½ bricks back from the first four. Step each successive three-course tier in this manner until the sheathing rises one brick higher than the landing.

2 Paving the steps. Mortar veneer to the first riser, using a mason's line to keep the veneer even with the projecting sides; make the top course even with the first tread of the core, cutting and adjusting as necessary. Then mortar veneer to the first tread, bringing the tread veneer flush with the riser veneer. Pave the remaining risers and treads in this manner, then cover the landing with bricks laid lengthwise.

Barbecues of Blocks and Bricks

Among the most practical and attractive applications of the technique of building with inexpensive blocks, then covering with handsome bricks, is a freestanding barbecue for the backyard. The bigger and more elaborate the barbecue, the more time and money you will save. In the basic barbecue illustrated below, 24 cinder blocks fill a space equivalent to 240 relatively costly bricks.

The purpose of such masonry is to support a cooking grill, a fire grate and an ashtray. All three are available from ma-sonry supply dealers, hardware stores and restaurant equipment houses. The grill, made of cast iron or steel, should be rigid enough not to sag under the weight of food, and heavy enough to resist being pushed out of position accidentally. The charcoal-supporting fire grates are usually made in small sections of heavy cast iron; you will probably need more than one. Or you can install a more so-phisticated grate-and-grill system—available assembled and ready to install—in which the chef adjusts cooking temperature by cranking the cooking grill or fire grate up and down. The ashtray can be anything from a specially built sheet-metal container to a disposable aluminum broiler pan.

You can adapt the construction shown on the following pages to build a barbecue to your special needs and tastes.

To change the height of a fixed grill, simply add or subtract courses of brick from the bottom of the casing and adjust the height of the core to fit. Or change the shape of the barbecue itself: Lay out the first course of the casing in the shape you want and plan a block core to fill the interior. In that way you can accommodate larger or smaller grills, increase or decrease the counter space, or build in a second grill.

The open space in the firebox below the ashtray can with little difficulty be modified to store charcoal or to serve as a warming oven. To do so, set into the front of the firebox a door that has a flange to support bricks above it; then build courses of brick up the sides and across the top of the door as high as the ashtray, and cover the space with a ¼-inch steel plate. The cover plate is set

A Plan for a Permanent Barbecue

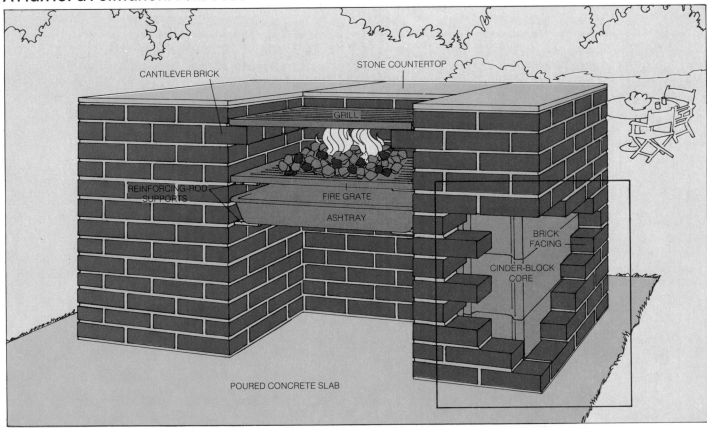

CANTILEVER BRICK

STONE COUNTERTOP

GRILL

REINFORCING-ROD SUPPORTS

FIRE GRATE

ASHTRAY

BRICK FACING

CINDER-BLOCK CORE

POURED CONCRETE SLAB

Anatomy of a barbecue. Beneath the brick skin of any block-and-brick barbecue lie courses of cinder block—one course of block for every three courses of brick—all resting on a 6-inch-thick concrete slab. Twelve brick courses laid with ½-inch mortar joints will position the cooking grill about the same height as burners on a kitchen stove.

The grill should be 6 to 9 inches above the fire grate. It is most attractively held by projecting cantilever bricks, while convenient supports for fire grate and ashtray are stubs of reinforcing rod fixed in mortar joints. These structural features, as well as the vertical dimensions, are common to most barbecues and are illustrated in the basic barbecue above. It measures about 5½ feet by 2½ feet, and fits on a slab 6½ feet wide and 5 feet deep. These horizontal dimensions can be varied to create other designs, and special conveniences —an oven or storage chamber—are easily added as explained in the text above so that you can make a barbecue to suit your own taste.

into a mortar joint around the firebox as the bricks are laid and used as a base for a shelf of bricks.

A more elaborate variation is an ash-removal and fire-stoking door equipped with a vent to control the draft. Install the door in a brick facing that closes in the firebox all the way to the top of the barbecue. As a final touch, you can use the rim of bricks around the cooking grill to support a portable hood that converts the grill into a smoke oven.

Before building any barbecue, choose a site with care. The prevailing summer breezes should be your first consideration. If possible, locate the barbecue downwind from your house, your neighbor's house and the dining area. Orient the barbecue so that the wind blows smoke away from the cook and creates a draft for the fire. Before starting construction, check with your local building inspector and fire department to see if local codes or permit requirements affect your plan.

The construction job itself falls into two distinct stages: pouring a reinforced concrete slab *(pages 58-60)*, and laying the brick and cinder block. Plan a slab large enough to provide a 6-inch skirt around the back and sides of the barbecue and a 2-foot apron in front for the cook to stand on. The slab should be 6 inches thick, poured onto 6 inches of gravel and reinforced with wire mesh. To estimate material requirements, adapt as necessary the list at right, which is designed for the specific barbecue illustrated on the opposite page. Stone for the top of the barbecue should be ordered cut to size from a mason supplier after the bricks are laid.

A Basic Shopping List

For the slab:
Gravel, 17 cubic feet
Concrete, 17 cubic feet
6″ × 6″, 10 gauge reinforcing mesh, 5′ × 6½′

For the barbecue:
15 cinder blocks, 8″ × 16″ × 8″
15 cinder blocks, 4″ × 16″ × 8″
400 SW bricks, 3¾″ × 2¼″ × 8″
Masonry cement, 350 pounds
Sand, 900 pounds
80 wall ties
⅜″ reinforcing rod, 20′
Lightweight cardboard (from a stationer)
⅜″ stone, 10 square feet (approximately)
Resin-based sealing compound for stone

Barbecue hardware:
Cooking grill
Fire grate
Ashtray

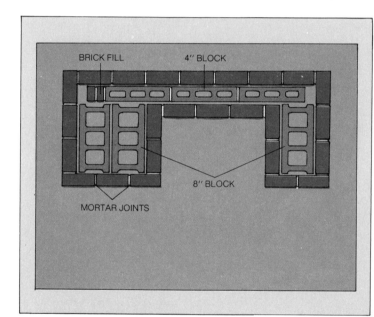

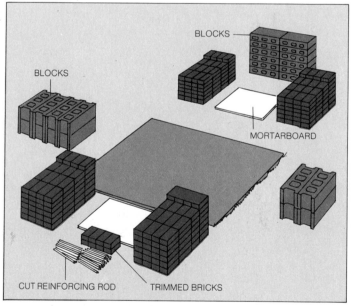

The pattern of bricks and blocks. Lay out the first course of brick in a dry run. Then, fit a block core inside the brick, filling leftover spaces with bricks or trimmed blocks. The casing bricks are not mortared to the blocks or the individual block walls to each other; however, bricks used to fill out the core are mortared to the block wall they complete *(left in drawing above)*. Bricks are laid in the running-bond style, with staggered vertical joints, but the blocks can be laid directly on top of one another. If you build the basic barbecue laid out above, a dry run is unnecessary; you can follow the plan exactly as drawn. In this case the 4-inch blocks at the back of the barbecue are laid in running-bond pattern, and two vertical bricks complete each course at alternate ends.

Setting up the work site. After the concrete slab is ready, save yourself hundreds of steps during the job by distributing bricks, blocks and mortar boards around the slab. Stack equal piles of bricks at the corners of the slab, about 2 feet from where you will be working; between the stacks, place mortarboards. Position cinder blocks near where they will be used. Before mixing mortar, trim blocks to fit the core if necessary. In addition, trim bricks to use as cantilever supports for the grill, and cut reinforcing rod to 7½-inch lengths to support the fire grate and ashtray. The basic barbecue opposite calls for whole, untrimmed blocks, but it requires eight trimmed bricks for the grill and 32 cut pieces of reinforcing rod for the grate and ashtray.

Putting Bricks and Blocks Together

1 Starting the brick casing. After a concrete slab has been poured and cured, use wall-building techniques to lay bricks. With a chalk line, mark a guideline for the back of the barbecue 6 inches from the edge of the slab. Using a steel square as a guide, chalk a second line 6 inches from one side of the slab, and build a stepped three-brick corner, or lead, at the intersection of the two lines. Extend the first course of bricks along the back of the barbecue to the far corner (in the basic barbecue shown here, lay five more bricks), then turn the corner at the other end, using the steel square again to make a true right angle. Construct a second three-brick lead at this corner, measuring carefully with a story pole.

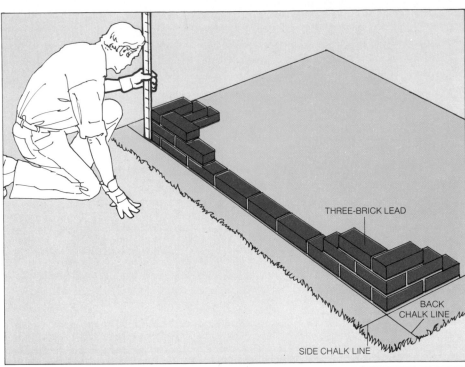

THREE-BRICK LEAD

BACK CHALK LINE

SIDE CHALK LINE

2 Starting the block core. Complete the first course of brick all around the barbecue, then lay the second and third courses along the sides and back. This leaves a hollow space into which the blocks will be lowered. In the basic barbecue illustrated, 4-inch blocks form the back core and 8-inch blocks form the side cores.

Trowel mortar onto the concrete slab, enough to make a bed for one block at a time. If you find that the mortar is so thin that it oozes out under the block's weight, sprinkle the bed with dry mortar to stiffen it. Space the blocks evenly inside the surrounding bricks without disturbing the bricks, leaving a ½-inch gap between bricks and blocks. Level the tops of the blocks with the top of the third-brick course by adjusting the thickness of the mortar beds under the blocks. To fill a course of blocks at its end, you may have to mortar in upended bricks.

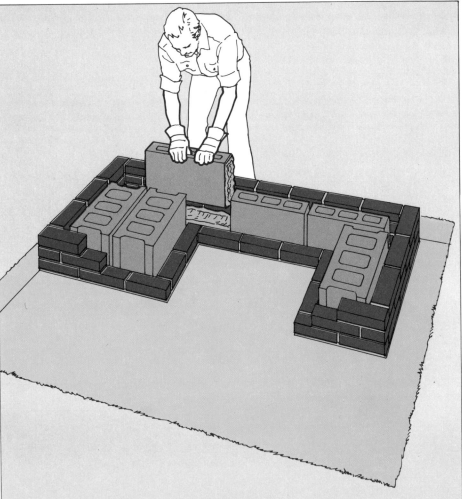

3 **Installing wall ties.** Complete the second and third courses of the brick casing, then lay wall ties across the gaps between blocks and bricks about every 10 inches, using diagonal ties at the corners. Throw the mortar bed for a three-brick lead at a rear corner of the barbecue, mortaring the wall ties in position as you proceed. When you come to a place for a tie, pick up the tie, lay the mortar, furrow the mortar with the trowel and push the tie into the mortar.

Repeat Steps 1 and 2. Install wall ties on every course of cinder block. When you lay the bed for a new course of block, bend the tie projections out of the way, trowel on the mortar, then bend the ties back into the mortar.

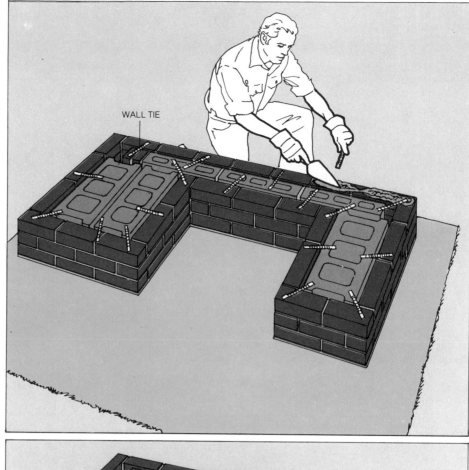

WALL TIE

4 **Setting supports for the ashtray.** Lay blocks and bricks until the block core is three courses high, and the brick casing rises nine courses along the sides and back but only seven courses around the firebox. As you lay the eighth course of fire-box brick, push 7½-inch lengths of precut reinforcing rod into the mortar bed on each side of the firebox. The rods should be evenly spaced and should project 4 inches from the casing. Lay the ninth course of bricks; then, in the mortar bed for the tenth course, set wall ties and reinforcing-rod supports for the fire grate.

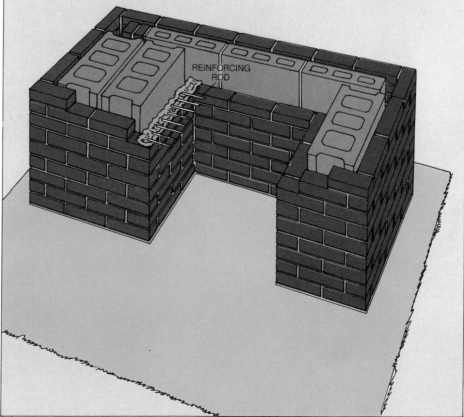

REINFORCING ROD

5 **Cantilevering bricks for the cooking grill.** As you lay the 12th course of brick, incorporate the trimmed bricks that will project, or cantilever, from the inner walls of the fire well, like brackets for a shelf, to support the cooking grill. Butter the side of each brick and lay it into the mortar bed so that the untrimmed end projects 4 inches. Set upended bricks on the inner edges of these bricks to keep them from toppling over while the mortar is setting.

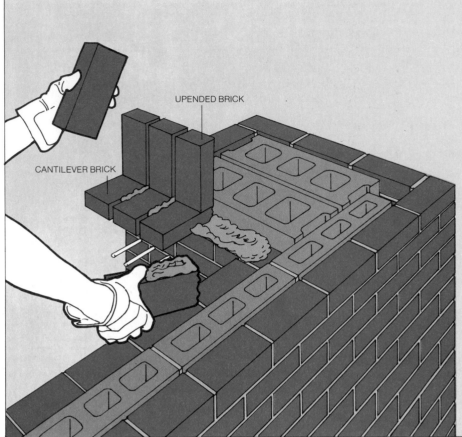

CANTILEVER BRICK

UPENDED BRICK

6 **Trimming the joints.** Let the mortar around the cantilevered bricks set for about 20 minutes. When the mortar is firm, remove the upended bricks and, with a ½-inch joint filler, trim excess mortar from the joints. Position the last set of wall ties and then lay the 13th course of brick around the top of the barbecue.

7 **Paving over the blocks.** Fill in the space on top of the fourth course of cinder block with a layer of brick in the pattern shown at right. Lay a mortar bed on the blocks and butter these bricks on the ends and one side, so the bricks are mortared to the cinder block, the inner sides of the brick casing and the adjacent topping brick.

8 **Topping the barbecue with stone.** Lay a large piece of cardboard over one side of the barbecue top, weighting the cardboard with bricks to keep it from shifting. Mark the exposed side to identify it as the top and draw as much of the outline of the barbecue as you can on the underside. Remove the cardboard, complete the outline with a straightedge, then cut along the lines to make a template and set it on the barbecue. Trim the template for a close fit, and make similar templates for the other side and center section. Trim ½ inch from each end of the center template to allow for mortar joints in the stone top.

At your mason-supply yard have three pieces of stone cut to match the templates. Lay them with mortar atop the barbecue, as shown in the instructions for a cinder-block wall. Let the mortar dry for two days, then coat the stone and its mortar joints with a resin-based stone-sealing compound, sold under such tradenames as Watco, to protect it against splatters of grease.

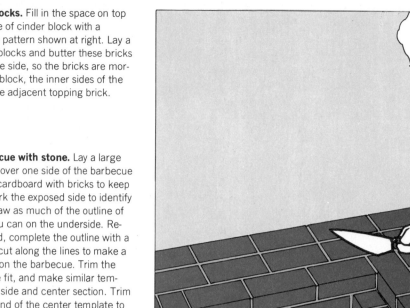

Picture Credits

The sources for the illustrations in this book are shown below. Credits for the pictures from left to right are separated by semicolons, from top to bottom by dashes.
Cover—Ken Kay. 6—Ken Kay. 10, 11—Al Freni. 12 through 17—Drawings by Whitman Studio, Inc. 18, 19—Drawings by Roger Metcalf. 20 through 25—Drawings by Nick Fasciano. 27, 28, 29—Drawings by Peter McGinn. 30 through 33—Drawings by Adolph E. Brotman. 34, 35—Drawings by Peter McGinn. 36—Drawings by Whitman Studio, Inc. 38—Ken Kay. 41—Drawings by Peter McGinn. 42 through 53—Drawings by Ray Skibinski. 54—Drawings by Peter McGinn; Enrico Ferorelli. 55—Dan Budnik—Enrico Ferorelli; drawings by Peter McGinn. 56—Courtesy Portland Cement Association. 57—Courtesy Portland Cement Association (2)—S. W. Newbury, courtesy of Concrete Construction, Addison, Illinois (2). 58 through 63—Drawings by Peter McGinn. 65, 66—Drawings by Ray Skibinski. 67—Drawing by Ray Skibinski—map courtesy U.S. Weather Bureau. 68, 69—Drawings by Adolph E. Brotman. 70 through 73—Drawings by Nick Fasciano. 74—Wolf von dem Bussche. 76—Drawings by Peter McGinn. 77—Drawings by Peter McGinn; Ken Kay. 78, 79—Drawings by Peter McGinn. 80 through 95—Drawings by Nick Fasciano. 96 through 101—Drawings by Ray Skibinski. 102 through 111—Drawings by Nick Fasciano. 112 through 117—Drawings by Whitman Studio, Inc. 118 through 123—Drawings by Ed Vebell.

Acknowledgments

The index/glossary for this book was prepared by Mel Ingber. The editors also thank the following: Abbey-Rent All, Hicksville, New York; Katherine M. Anderson, Concrete Construction Publications, Inc., Addison, Illinois; R. B. Becker, Marshalltown Trowel Co., Marshalltown, Iowa; Don Boyce, Don Boyce Studios, New York City; Building Officials and Code Administrators International, Chicago; Bruce Burkland and Daniel C. Cammer, Brick Institute of America, McLean, Virginia; Henry B. Comstock, Blauvelt, New York; Ed Crawford, Manzo Associates, Inc., Turnersville, New Jersey; Clyde Epifanio and Milt Stern, Nemo Tile Supplies, Jamaica, New York; Warren Fenske, Goldblatt Tool Co., Kansas City, Kansas; Forbes-Ergas Design Associates, Inc., New York City; Franco Bros., Greenwich, Connecticut; Richard D. Gaynor, National Ready Mixed Concrete Assoc., Silver Spring, Maryland; Huey Hooter, Construction & Fastener Products, Desa Industries, Park Forest, Illinois; International Conference of Building Officials, Whittier, California; Robert Kleinhans, Tile Council of America, Princeton, New Jersey; A. B. Miller, The Belden-Stark Brick Corp., New York City; National Concrete Masonry Assoc., McLean, Virginia; Frank Randall, Portland Cement Association, Skokie, Illinois; Buck Richardson, International Masonry Institute, Washington, D.C.; W. H. Silvers Hardware, Inc., New York City; Southern Building Code Congress International, Inc., Birmingham, Alabama; Carl Starner, Black & Decker Manufacturing Co., Towson, Maryland; Stern & Hagmann Architects, New York City; Richard P. Wens, Red Devil Inc., Union, New Jersey.

Index/Glossary

Included in this index are definitions of many of the masonry terms used in this book. Page references in italics indicate an illustration of the subject mentioned.

Adhesives: epoxy, 34, 35; mastic glue, 34

Anchoring poles and posts, *30-33;* concrete mixture, 30; flagpole hinges, *32;* flagpole sleeves, *31;* playground gym, 33

Anchors, expansion, 34

Asphalt: *a concrete with a petroleum product as binder.* Repairing cracks, 25; repairing holes, *25;* sealer

Asphalt, cold mix: *material used to patch large holes in asphalt.* Types, 25; use, 25

Barbecue, *110;* block core, *120-121;* brick and block pattern, *118,* 119; casing for, 112, *118, 120-123;* concrete base, 119; hardware, 118, 119; laying brick, *120;* materials, 119; reinforcing rods, 119, *121;* site, 119; stacking materials, *119;* stone top, 119, 123; variations, 118-119; wall ties, 112, *121*

Bat: *segment of a cut brick.* Definition, 9; making, *17*

Batter board: *supports for strings used to locate the edges of a proposed structure or excavation.* Construction, 66

Bed: *largest face of a brick.* Described, 77

Bed: *layer of mortar on which blocks are laid.* For paving, *84;* preparing, 13; for tile, *96, 97*

Bird bath: *depression in concrete slab.* Eliminating, 52

Blocks, 75; building with, 75; esthetic effect, 75. *See also* Brick; Concrete block; Stone; Tile

Bluestone, 102; for dry walls, 106

Bond, pattern: *pattern of laid masonry blocks.* Variations for brick pavement, *78-79;* for wall, 86

Brick: *baked clay block.* Building, 76; buttering, *14;* cantilevering, *122;* cleaning, 37; closure, *14;* cored, *76;* estimating quantity, 76; face, 76; fasteners for, 34; paving, *74, 76;* paving with, *78-85;* removing, 18; repairing, 14, 16; replacing, *18;* sizes, 76; splitting, *17;* storing, 76; surface names, *77;* surface textures, *77;* veneer

or casing for concrete block, 112, 116, *117*

Brick cube: *prepackaged set of bricks.* Described, 76

Brick projects: barbecue casing, 112, *118;* estimating material, 86; path, *78-85;* patio, *78-85;* planning, 86, 87; veneer, 112, 116, *117;* wall, *86-95*

Bricklaying, *14;* finishing joints, *15;* for freestanding wall, *88-92;* for wall with corners, *94-95. See also* Paving

Brickset: *broadbladed chisel for splitting brick.* Described, 10, *11;* use, *17, 18*

Building codes, 8, 9, 40; for footings, 65; and reinforcing concrete, 58

Butter, cement: *slightly wet cement-water mixture.* Use, 102, *105*

Buttering: *applying mortar to a surface.* Described, 9, *14*

Caulk, for tile expansion joints, *101*

Cement, air-entrained portland: *cement that produces tiny air bubbles in concrete.* Described, 26

Cement, portland: *bonding agent in mortar, grout and concrete.* Basic to masonry, 8; butter, 102, 105; hardened bags, 26; pigmented, 12, 27; size of bags available, 26

Cement, masonry: *mixture of portland cement and hydrated lime.* Proportion in mortar, 12

Cement paint: *thin mixture of cement and water.* Use, *21*

Chalkline: *cord used to mark straight lines for masonry.* Use, *88*

Chisel, cold, 10; use, 20. *See also* Brickset

Chisel, stonemason's, 10, *11*

Cinder block: *lightweight concrete block. See* Concrete block

Cleaning: of leftover mortar and concrete, 37; of stained or blemished masonry, 37; of tools, 37

Clothesline poles, anchoring, 30

Coarse aggregate: *small stones that provide bulk in concrete.* Function, 26; proper for concrete, 26

Coloring masonry: concrete, 27, 50, 51; dry-shake, 50, 52; mortar, 12, 24; with paint, 27, 50

Concrete: *a mixture of cement, filler and water, which hardens to a rocklike mass.* Bleed water, 50, 52; breaking up, 40; cleaning, 37; coloring, 27, 50; creep, 58; curing, 50, *53;* fasteners for, 34; faults, *56-57;* finishing surface, 50,

54-55; forms, 39; ingredients, 26; mixing by hand, 26, *27,28;* mixing by machine, 26, *27, 29;* mixture for spreading, 26; pouring for footings, 65; pouring into slab forms, *50-53;* ready-mix, 40, 41; recipes, 26; reinforcement, 39; repairing, 16; repairing steps, *22-23;* repairing stucco, *24;* repairing wide cracks, *20-21;* sealing, 56, 57, 70; spalled, 21; speed of working, 39, 50; storing ingredients, *27;* tensile strength, 39; testing consistency, *29,* 40. *See also* Forms for concrete

Concrete, air-entrained: *concrete with tiny air bubbles, highly weather resistant and easily workable.* Bleed water, 50; described, 26; mixing, 26

Concrete, reinforced: *concrete embedded with steel mesh or rods to add tensile strength.* Breaking up, 40; purpose, 39; wire mesh, 58, *60*

Concrete block, 74; cinder block, 112; cleaning, 37; coping, *113;* cost, 112; fasteners for, 34; laying, *113;* mortar for, 112; rainproofing, *113;* screen, *112, 114;* shapes and sizes, 112; veneered or encased, 112; webs, 112

Concrete block projects: barbecue, 112, *118-123;* retaining wall, *114-115;* staircase, *116-117;* walls, 112, *113-114*

Concrete projects: clearing site, 40; driveways, *62-63;* estimating materials, 40; grading, *45-47;* ground stability, 40; patio, *58-61;* planning, 40; pond or pool, *70-73;* pouring concrete into forms, *50-53;* ready-mix delivery, 40, *41;* reinforced slabs, *58-61;* sidewalks, *42-53;* steps, *68-69;* texturing surface, *54-55*

Concrete slabs: as base for barbecue, *119;* as base for brick pavement, 80, *84;* as base for concrete block steps, 116; as base for stone pavement, *105;* building forms, *42, 45, 47-49;* coloring, 50, 51, 52; control joints, *38, 39, 50, 53;* curing, 53; for driveways, *62-63;* edging, 50, *53;* estimating materials, 40; expansion joints, 49; with exposed aggregate, *55;* extending, 64; faults, *56-57;* finishing, 50, *52-53;* grading for, *45;* joint filler, 49; with permanent forms, *61;* pitch, 42, 45; pouring, *50-51;* preparing base, 40, *42, 43-45;* reinforced, 42, *58-61;* textured finishes, *54-55;* veneering with tile, *96-101*

Cracked concrete, described, *57. See also* Repairs

Crazing: *fine cracks in concrete.* Described, *57*

Curing: *hardening of cement mixtures when kept moist.* Concrete, 50, *53,* 56; of epoxy mortars, 20; and moisture, 16, 20

Darby: *trowel used to compact and level poured concrete.* Described, 9, *10-11;* use, 50, *52;* used to embed stones in concrete, *55*

Drainage beds, 40; installing, *49*

Driveway, concrete, 58, 62, 64; excavating and grading, *63;* finishing, *63;* pitch, 58, 62; wire mesh, 58, 62

Dry-shake: *powdered coloring mixture for concrete.* Use, 50

Dusting: *powdering of concrete surface.* Described, *56*

Edgers: *tools used to shape edges of concrete.* For pond, 73; shown, *10;* step, 10; use on steps, 50, *53*

Efflorescence: *white powdery coating produced on masonry by salts rising to the surface.* Removal, 37

Epoxy: *strong, durable adhesive.* Used to fasten studs in masonry, 34, *35*

Epoxy compound mortars, used to repair concrete, 20, 21, *22*

Fasteners for masonry, 34-36; driving nails, *36;* installing studs, *35;* making holes, 34-35

Fence posts, anchoring, *30*

Fieldstone, *74;* shown, *102;* for walls, 106-109

Flagpole, anchoring, 30; hinge, 30, *32;* sleeve, 30, *31*

Flagstone: *stone split into slabs.* Uses, *102;* concrete imitation, *55;* estimating quantity, 102; laid dry, 102; mortared paving, *105;* mortarless paving, *103-104;* thickness and bed type, 102

Float: *wooden trowel or board used to finish concrete surfaces.* Bull float, *60;* shown, *10;* use on concrete slabs, *52;* use on stucco, *24*

Footing, concrete: *separate part of structure that rests directly upon the ground.* Concrete mixture, 65; depth, *67;* estimating materials, 40; excavation for, 40; forms for, 65; for heavy structure, 65, 66-67; for heavy wall, 86, 87, 88; for heavy wall with corners, 93; for low, light wall, *65;* reinforced, 65, *67;* for stone wall, 110

Form oil: *lubricant for forms to prevent the forms from sticking to concrete.* Use, 68, 69

Forms for concrete: *bottomless boxes used to shape concrete.* Described, 39, 42; for footings, *65, 66-67;* materials, 42; metal, 42, 48, 49; permanent, 42, *61;* pouring into, *50-51;* for slabs, 42, 45, 47-49; stakes, 40, 45-47; for steps, *68, 69*

Frost line: *deepest penetration of frost below ground level.* Determination, 67

Furrowing mortar: *making a shallow depression in a line of mortar to help the mortar spread.* Described, 13

Gloves, use, 12

Glue, mastic, use, 34

Goggles, use, 16

Grade: *slope in a concrete surface for drainage.* Lengthwise, *45;* widthwise, 46-47

Granite, 64, 102

Gravel, drainage bed, 40, *49. See also* Coarse aggregate

Green mortar: *set, but not thoroughly hardened mortar.* Definition, 9

Grout: *mortar mixed for especially thin consistency.* Curing, 16; used for brick paving, *85;* used for repairs, 16, 19; used for stone paving, *105*

Grout, tile, 96; application, 96, *100-101*

Hammer, ball-peen: to split brick, 17

Hammer, bricklayer's: *hammer with one chisel-shaped end for cutting or trimming brick.* Shown, *11;* used for splitting brick, 17

Hammer, stonemason's: described, 10, *11;* use, *103*

Hammer drill: *power drill producing repeated impacts of bit.* Use, 34, 35, 64

Hawk: *a small mortarboard with a handle.* Described, 10, *11;* use, *16;* used to replace a brick, *18*

Header: *smallest face of a brick.* Described, *77*

Header course: *row of bricks with headers exposed.* Buttering, 14; described, *77*

Hod: *v-shaped trough used to carry bricks or mortar.* Described, 9

Hoe: *long-handled tool used to mix concrete or mortar.* Shown, *10-11;* use, 12, 28

Holes in masonry, making, 34; locating, *34;* tools, 34

Holes for posts, digging, *30*

Joint, control: *joint designed to be the focus of stresses in concrete.* Making, *38, 39,* 50, *53*

Joint, expansion: *joint between concrete slabs to allow for expansion.* Filling when tiled, *101;* installing filler, *49*

Joint, mortar: *mortar connecting blocks, bricks or stones.* Indoor and outdoor, 14, 15; jack, *92;* making different types, 15

Joint filler: *flexible material used in concrete slab joints to allow for expansion.* Installation, *49*

Joint filler: *tool for filling long joints with mortar.* Described, 10, *11;* use, *16*

Joint reinforcement: *metal lattice laid in joints to strengthen bonds.* Installing, *114*

Jointer: *tool used to cut grooves in fresh concrete.* Described, *10-11;* used, *38, 39,* 53

Jointers, concave and v-shaped: *tools used to shape profile of mortar joints.* Described, 10, *11;* use, 15; used to give concrete a flagstone pattern, 55; used in paving, *85*

Leads: *first courses at ends of brick, block or stone walls, which define length and alignment.* Making, *88-90, 91;* making corner leads, *94-95*

Level, mason's: *long level used to check alignment of work.* Shown, *10;* use, 14

Lime, hydrated: *ingredient of mortar.* Proportions in mortars, 12

Limestone, 102; for dry walls, 106

Mallet, shown, *10;* use, 17

Masonry, advantages, 7; blocks, 75; cleaning stained or blemished, 37; weaknesses, 7

Masonry bit: *carbide-tipped drill for masonry.* Described, 10, *11;* use, 34, 64

Masonry-core bit: *bit that cuts cylindrical plug or hole in masonry.* Removal of remaining plug, *35;* use, 34, *35*

Mason's line and blocks: *device used to lay true courses of masonry.* Described, 10, *11, 90;* use, *90, 91*

Maul: *heavy-headed hammer.* Use, *35*

Measuring tape, *10*

Mixer, power: *motorized device for mixing concrete.* Cleaning, 37; with pigmented cement, 27; use, 26, *29*

Mortar: *cement-based material used to bond masonry blocks and make repairs in concrete.* Applying bed, *13*; applying to brick, *14*; cleaning up leftovers, 37; colored layer on concrete, 27, 50, 51; coloring, 12, 24; curing, 16; described, 12; fasteners for, 34, 36; finishing joints, *15*; furrowing, *13*; green, 9; handling, 12; mixing, *12*; ready-mix, 12; recipes, 12; repairing with, *16*; sand-cement-epoxy, 16, 20; throwing, *13*; for tile, 96, *97*; used for brick paving, 76, 78, 80, 84, *85*; used for concrete block, 112, *113*; used for stone paving, 105; used for stone walls, 110

Mortarboard: *board for holding mixed mortar.* Use, *13*

Mortar rake: *tool used to remove mortar.* Use, *18*

Moss, removal, 37

Mud: *wet concrete or mortar.* Defined, 9

Nails, masonry: *hardened steel nails.* Driving, *36*; use, 34, 36

Oil and grease, removal, 37

Paint, removal, 37

Path, brick: cored brick for, 76; patterns, *78-79*; paving, *80-85*

Path, tile, *96-101*

Patio, brick. *See* Path, brick

Patio, concrete: excavating and grading, 58, *59*; extending, 64; pouring and finishing, *60-61*; slabs, 58

Patio, tile, *96-101*

Patterns of masonry, brick, *78-79*

Paving, brick, 76, 78; for concrete-block steps, 116, *117*; edging, *80-81*, 84; and frost, 78, 80; mortared, 76, 80, *84-85*; patterns, *78-79*; unmortared, 76, *80-83*; sand bed, *82. See also* Concrete slabs

Paving, stone, 102; mortared, *105*; mortarless, 102, *103-104*

Playground equipment, anchoring, 30, *33*

Plumb bob: *device for assuring verticality.* Used with batter boards, 66; for flagpole sleeves, *31*

Pointing: *replacing of old mortar.* Described, 9, 16

Polyethylene rope: *joint filler for tile.* Described, 96; use, *101*

Pond, concrete, *70*; concrete mixture, 70; draining, 73; excavating, *71*; fish for, 70,73; laying in mesh, 71; pouring and finishing, *72-73*; runoff point, 70, *73*; sealing, 70; special tools, 70

Pool. *See* Pond, concrete

Pop outs: *holes in concrete where aggregate has come loose.* Described, 57

Posts, anchoring, *30-33*; fence posts, *30*; flagpole hinge, *32*; flagpole sleeve, *31*; preservatives, 30

Raking: *removing uncured mortar from joints for decorative effect.* Brick joints, 15; definition, 9; flagging joints, 105; stone wall joints, 110, *111*

Raking tool: *tool used to remove mortar from joints.* Described, 10, *11*; use, 15

Rebar. *See* Reinforcing rod

Reinforcing rod: *metal rod embedded in concrete to increase its tensile strength.* Binding slabs, 64; in barbecue, 119; in footings, 67

Repairs: to asphalt, *25*; to concrete steps, sidewalks and driveways, 16, *20-23*; to mortar, 16; to stucco, 24; to walls, *16-19,24*

Retempering: *adding water to unused mortar to prevent drying.* Described, 12

Right angles, determining, 66

Rub brick: *tool used to smooth a slab for a tile covering.* Shown, 10; used, 97

Rubble: *uncut stone.* For dry walls, 106; quarried, 102, 106

Rust, removal, 37

Safety procedures: eye protection, 16; handling cleaning chemicals, 37; handling mortar, 12; handling stone, 102; handling wire mesh, 58

Sailor: *brick standing upright with broad face showing.* Course, 77; definition, 9; as pavement edging, 81

Sand, as fine filler in mortar and concrete, 26; proper type for concrete, 26; proportions in mortars, 12; types, 12

Sand bed, for paving, 80, 82

Sandstone, for dry walls, 106

Screed: *long board used for leveling surface.* Defined, 9; use on concrete slabs, 50, *51*; use on gravel bed, *49*

Sealer, concrete: *compound used to seal pores in concrete.* For pond, 70; use, 56, 57

Slate, *74*, 102

Slope gauge: *tool for measuring taper of stone walls.* Constructing, *106*; use, 108

Smoke, removal, 37

Soldier: *brick standing upright with narrow face exposed.* Course, 77; defined, 9; as pavement edging, 81

Soup: *runny concrete.* Defined, 9

Spalled concrete: *concrete with flaked surface.* Described, 56; repairing, 20, 21

Square, steel: *L-shaped tool used to check right angles.* Shown, 10

Stains, removal, 37

Star drill: *chisel with X-shaped end used to make holes in masonry.* Described, 10, *11*; use, 34

Steps, repairing, 22-23

Steps, concrete: dimensions, 68; estimating, materials, 40; pouring and finishing, 69; profile, 68

Steps, concrete block: dimensions, 68, 116; drainage, 116; filling shell with earth, 116; laying block, 116; veneering, 116, *117*

Stone, 102; cleaning, 37; estimating quantities, 102; fasteners for, 34; laid dry,102,*106-109;*sealing compound,119, 123; suited to dry walls, 106; suited to wet walls, 110; types, *102*

Stone, building with, 102; barbecue top, 119, 123; dry walls, 102, *106-109*; estimating materials, 102; mortared paving, *105*; mortarless paving, *103-104*; slope gauge, 106, *107*; safety precautions, 102; wet walls, *110-111*

Stone, square-cut: *roughly shaped stone.* Ashlar, 102; described, *102*

Story pole: *gauge used to check height of a course of masonry.* Defined, 9; making, *86*; use, 88

Stretcher: *long, narrow side of a brick.* Described, *77*

Stretcher course: *row of bricks with stretchers exposed.* Buttering, *14*; described, 77; rowlock, 77

Strike-off board. *See* Screed

Stucco: *mortar containing hydrated lime, used for veneering.* Mortar for patching, 24; repairing, 16, *24*

Stud: *strong fastener formed by a bolt epoxied into masonry.* Fastening to masonry, *35*; use, 34

Stud driver: *gun used to drive nails with gunpowder.* Renting, 8; use, *36*

Tamper: *heavy block of wood used to pound and compress layers of concrete or asphalt.* Compacting base for concrete, *44;* curved, *70, 71;* use with asphalt, *25*
Tar, removal, 37
Textures, brick, 77
Throwing mortar: *depositing mortar from trowel.* Described, *13*
Tile: *thin slab of baked clay.* Cleaning, *101;* cutting, *99;* edging, *101;* estimating material, 96; joints, 96, *100-101;* laying, *98, 100;* mortar bed, 96, *97;* ripper, 10, 11; unglazed, 96; water resistance, 96; veneering concrete with, *96-101*
Tile, mosaic, 96; laying, *100*
Tile, paver, 96; cutting, *99;* laying, *98, 100-101;* veneering with, *96-101*
Tile, quarry, 74, 96; cutting, *99;* laying, *98, 100-101;* veneering with, *96-101*
Tile cutter, 10, *11;* use, 96, *99*
Toggle anchor, 34
Tools, *10-11;* cleaning, 37
Transit-mix concrete: *concrete power-mixed in a delivery truck.* For slabs, 50; unloading, *41;* use of, 41
Trowel, box notch: *square-notched trowel used to score bed joint for mosaic tile.* Described, 10, *11*
Trowel, mason's: *tool for spreading mortar.* Described, 10, *11;* parts, *13;* square-end, 10; technique for making mortar bed, *13;* used for shaping joints, 15
Trowel, pointing: *small triangular trowel used for shaping joints.* Described, 10, *11;* use, 16
Trowel, rectangular notched: *trowel used to score bed joint for tile.* Described, 10, *11;* use, 97
Trowel, V-notched: *square trowel used to score bed joint.* Described, 10, *11,*

Wall tie: *metal strips used to link masonry tiers for reinforcement.* Connecting casing to core, 112, *121;* described, 87, *89;* installed, *89, 91*
Walls, brick, 86, 87; basic bricklaying, *14-15;* brick proportions, 86; controlling heights of courses, 86; cored brick for, 76; estimating material, 86; footing, 65; height, 91; laying brick, *88-92;* leads, *88-90, 91;* pattern bonds, 86; planning, 86, 87; repairing, *16-19;* top rowlock course, 86, 87, 92; wall ties, *89,* 91
Walls, concrete block: coping, *113;* cost, 112; footing, 112; height, 112; laying blocks, *113;* laying screen blocks, *114;* pilasters, *114, 115;* rainproofing, *113;* reinforcement, *114;* retaining wall, *114-115;* stucco, *64B;* weep holes, 112, 115
Walls, dry stone, *106;* bed, *106;* bonding stone, *106, 107;* corners, *109;* laying stone, 106, *107;* sealing with mortar, 106, *109;* shape, *106;* shimming and chinking, 106, *108;* stone for, 106
Walls, wet stone, 106; bonding stone, *111;* concrete footing, *110;* drainage, 110; joints, 110, *111;* laying stone, *111;* shape, 110; stone for, 110; weep holes, 110, *111*
Walls. *See also* Stucco
Walls with corners, brick: bricklaying, *94-95;* footings, *93;* rowlock course, *95*
Water: amount required in concrete, 26; amount required in mortar, 12; bleed, 50, 52
Webs: *partitions in hollow core of cement or cinder blocks.* Described, 112; mortaring, *113*
Wheelbarrow, use, *28, 29*
Windrow: *wedge of mortar on trowel.* Described, 13
Wire mesh, reinforcing: bending for bowl, 70, *71;* laying for slabs, *60;* size, 58; use, 58
Works brick: *layout needing no block or brick cutting.* Defined, 9

Yard: *cubic yard.* Defined, 9